OXFORD BIOLOGY PRIMERS

Discover more in the series at
www.oxfordtextbooks.co.uk/obp

Published in partnership with the Royal Society of Biology

EVOLUTION

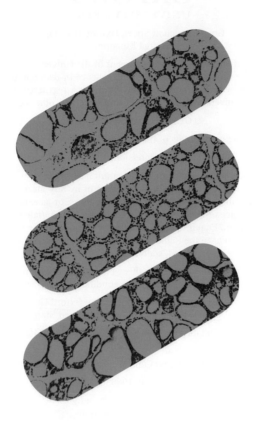

OXFORD BIOLOGY PRIMERS

EVOLUTION

Neil Ingram
Sylvia Hixson Andrews
Jane Still

Edited by Ann Fullick
Editorial board: Ian Harvey, Gill Hickman, and Sue Howarth

OXFORD
UNIVERSITY PRESS

Royal Society of
Biology

OXFORD
UNIVERSITY PRESS

Great Clarendon Street, Oxford, OX2 6DP,
United Kingdom

Oxford University Press is a department of the University of Oxford.
It furthers the University's objective of excellence in research, scholarship,
and education by publishing worldwide. Oxford is a registered trade mark of
Oxford University Press in the UK and in certain other countries

Published in the United States of America by Oxford University Press
198 Madison Avenue, New York, NY 10016, United States of America

British Library Cataloguing in Publication Data
Data available

Library of Congress Control Number: 2021940144

ISBN 978–0–19–886257–4

Printed in Great Britain by
Bell & Bain Ltd., Glasgow

PREFACE

Welcome to the Oxford Biology Primers.

There has never been a more exciting time to be a biologist. Not only do we understand more about the biological world than ever before, but we're using that understanding in ever-more creative and valuable ways.

Our understanding of the way our genes work is being used to explore new ways to treat disease; our understanding of ecosystems is being used to explore more effective ways to protect the diversity of life on Earth; our understanding of plant science is being used to explore more sustainable ways to feed a growing human population.

The repeated use of the word 'explore' here is no accident. The study of biology is, at heart, an exploration. We have written the Oxford Biology Primers to encourage you to explore biology for yourself—to find out more about what scientists at the cutting edge of the subject are researching, and the biological problems they're trying to solve.

Throughout the series, we use a range of features to help you see topics from different perspectives.

Scientific approach panels help you understand a little more about 'how we know what we know'—that is, the research that has been carried out to reveal our current understanding of the science described in the text, and the methods and approaches scientists have used when carrying out that research.

Case studies explore how a particular concept is relevant to our everyday life, or provide an intimate picture of one aspect of the science described.

The bigger picture panels help you think about some of the issues and challenges associated with the topic under discussion—for example, ethical considerations, or wider impacts on society.

More than anything, however, we hope this series will reveal to you, its readers, that biology is awe-inspiring, both in its variety and its intricacy, and will drive you forward to explore the subject further for yourself.

ABOUT THE AUTHORS

Syliva Hixson Andrews BS BSc MSc PhD, chapters 5 and 6

Sylvia studied in the United States and came late to the study of biology. After six years in biomedical research at Loman Linda Medical University in California, she returned to university at the age of 26. Intending to get her BS in Biology, she took an 'extra' in Physical Anthropology and immediately saw that many of her interests in biology dovetailed with human evolution. She completed two degrees, one in biology and one in anthropology, and went on to do her Masters and PhD in Biological Anthropology at U.C. Berkeley in California. After many years teaching at Community Colleges, she married and moved to the UK. Family circumstances dictated that she could not take full-time academic work, but because of her studies in Biological Anthropology she found job opportunities in diverse fields such as product development at a brewery, and statistical analysis for Dorset Adult Education. Sylvia has continued her interests in anthropology and is now the Director of Blandford Museum in the town of Blandford Forum, Dorset. She has worked at fossil sites in the Siwaliks of Pakistan (Miocene apes), Laetoli in Tanzania (Pliocene hominins), and Azokh Cave, an archaeological site in Nagorno Karabakh (now Azerbaijan). Her training in and understanding of human biology and evolution has allowed her to work in a variety of fields, all of which she has enjoyed.

Neil R Ingram BSc(Hons) Msc PhD PGCE, chapters 3 and 4

As a child, Neil explored fields and hedgerows in the countryside around the industrial city where he lived. He became curious about ferns, fruits, and fungi, wondering what and why they were. This was the time of lavish colour TV documentaries like 'Life on Earth' that added wonder to his curiosity. Neil's PhD research, carried out at the University of Birmingham, UK, was in the early days of genetic engineering and involved transferring single genes from parents to offspring. He rediscovered Mendel by breaking his laws. Multiple measurements from tens of thousands of tobacco plants were shaped into patterns that he explained in the journal *Nature*. Neil wanted a career where he could inspire curiosity in others, so he completed a PGCE at the University of Cambridge, UK. For many years he taught biology, psychology, and ICT in secondary schools, including Clifton College, where he was Head of Science. More recently, he has been Senior Lecturer in the School of Education in the University of Bristol, teaching postgraduate students, especially those on initial teacher training courses. He admires Charles Darwin as a great storyteller and thinks that the living world cannot be reduced to chemistry, physics, and mathematics without destroying its essence.

Jane Still BSc(Hons) PGCE, chapters 1 and 2

Jane Still grew up on the bayous of the Gulf Coast of Texas, the daughter of a naturalist and a chemical engineer in the oil industry. NASA built facilities for Mission Control and astronaut training while she was still at primary school. Many of her friends' parents were astronauts and space scientists, and she discovered that ordinary people do extraordinary things. The local schools had competitive science and maths research fairs, and for one Jane mapped the death of vegetation along her bayou as the land sank due to oil and water extraction. She had decided to become a marine biologist when as a nine-year-old she met a man on her village beach looking for a rare species of clam. From that moment on she became actively interested in conservation. To this day she still organizes ecological events for the local community and for schools, and is a member of the British Ecological Society.

   After doing a biology degree at the University of Exeter and PGCE at the University of Cambridge, Jane has taught in a variety of schools and is an experienced online tutor, examiner, and author.

Consultant: Dr Peter Andrews, chapters 5 and 6

Peter Andrews has a PhD and ScD from the University of Cambridge, UK, and has spent most of his career in human and primate evolution. For 35 years he worked at the Natural History Museum, London, where he was Head of Human Origins until 2000. His work covers hominid evolution up to, and including, humans and he has also worked extensively on taphonomy. He is currently curator of Blandford Museum, while retaining an Emeritus Research Associate position at the Natural History Museum. Peter also has honorary professorships at University College London and the University of York. He has published ten books, two with Professor Chris Stringer FRS, and nearly 200 articles in the scientific and popular press.

Consultant: Professor Derek Briggs, chapters 2 and 3 (Burgess shales)

Professor Derek Briggs is the Yale University G. Evelyn Hutchinson Professor of Geology and Geophysics, Curator of Invertebrate Paleontology at Yale's Peabody Museum of Natural History, and former Director of the Peabody Museum. He has made several remarkable discoveries of exceptionally preserved fossils and his research on the arthropods from the Burgess Shale of British Columbia has helped to transform our understanding of life in the Cambrian period. He has pioneered a combination of new experimental approaches to investigating the processes involved in fossilization of the 'soft parts' of animals.

ACKNOWLEDGEMENTS

Neil Ingram

It has been a pleasure to work with Ann and my co-authors, who have helped to make this book a real pleasure and never a chore.

I would like to thank three professors. Firstly, Derek Briggs was very enthusiastic for this project, and I am grateful for his willingness to contribute to it. Ian Trueman first showed me liverworts and fern allies and invited me to think. The late John Jinks taught me to work with discipline and rigour.

My colleagues, friends and pupils over the years, especially at the University of Bristol, Clifton College, Pearson and ISEB have always been supportive and helpful. I could not ask for more.

My parents, Roy and Eileen taught me to value education, and I will forever be grateful to them for their support.

Last, but not least, to my family: Helen, Paul, Mark and Abi are always the brightest stars in my sky and I thank them for their patience and understanding as I rush to make yet more deadlines.

I hope you all enjoy this book.

Sylvia Hixson Andrews

Ideas are formed through discussion, and I have had the privilege of being involved in many discussions about the evolutionary process over many years and with many people, too many to count and too many to thank. But I would like to mention Mireille Giovanola for many hours of conversation on all subjects. My husband, Peter Andrews, has always been ready to listen and debate, not just about primate evolution and behaviour, but ecology, taphonomy, and any and all aspects of life on earth. He has been an enormous help, and I continue to learn so much from him. Finally, thank you to my daughter, Chloe, for support of a much appreciated and different kind.

I would also like to thank Ann Fullick for inviting me to participate in this project. She is amazing in how she has pulled all three authors together, and has been very patient and helpful through the process. I have learned a lot from her!

Jane Still

My home county Dorset is a fantastic place for fossils, and I am indebted to the time and enthusiasm of the owners of two Dorset collections: Steve Etches and Wolfgang Grulke. Both are self-taught experts. The Etches Collection in Kimmeridge is now housed in its own museum, and focuses on the local marine life in the Jurassic era, especially ammonites. Wolfgang's collection is the inspiration for his prize-winning *Deep Time Trilogy* of books—*Heteromorph*, *Nautilus*, and *Beyond Extinction*. Having seen around it, Sir David Attenborough wrote, "I am, truly, lost for words". Long before that were the influences from my parents: my mother Sylvia who taught me how to observe the natural world and reflect on its patterns, and my father John who mentored me as I did my own research for school science fairs. And finally I am grateful to my own family Jonathan, Harriet, and William for their patience when normal routine and conversation was annihilated by evolutionary musings!

CONTENTS

1 INTRODUCING EVOLUTION

How many different kinds of animals and plants are there in the world? A survey in 2011 suggested that there are about 7.8 million species of animals and nearly 300,000 species of plants. Astonishingly, nearly 90% of these species have yet to be discovered, described, or named. Many are found in inaccessible hotspots of biodiversity, such as tropical rain forests and the deep oceans. Given the rate of man-made habitat destruction, it is possible that many of these species will become extinct before they can be discovered by scientists (see Figure 1.1). But these 8.1 million species are, for now, the success stories of evolution. Each is a unique way of solving the problems of surviving and reproducing in an unforgiving and changing environment.

How did so many species come to exist on Earth? The weight of scientific evidence suggests they all developed from a single common ancestral organism that lived about 4 billion years ago. This organism, imaginatively named LUCA (Last Universal Common Ancestor), is not thought to be the first organism on Earth, but whilst the others became extinct, LUCA's offspring went on to populate the whole planet. In doing so, they developed into new species that looked very different to the original LUCA, whilst sharing the same genetic code and many life processes.

New species develop from existing ones and may eventually replace them. The birth and death of species is central to the evolution of life on Earth.

Figure 1.1 Fire can destroy habitats and species—it is one of many factors which affect evolution and extinction around the globe.

This book explains how species form and change, using the latest scientific evidence. It tells the story of evolution, of the scientists who discovered the hidden mechanisms that lead to the birth and death of species. But above all, it celebrates the survival of the descendants of LUCA.

Introducing evolution: learning to think in a new way

Evolution is sometimes described as a 'theory', which some people take to mean that it is a speculative, untested, set of ideas. Nothing could be further from the truth. The same set of simple principles has been shown, time and again, to account for the emergence of all kinds of life, from bacteria and blue whales to humans. But this understanding of where we all came from has taken a long time to build.

The rise of English natural history

Our understanding of the scientific basis of evolution owes much to the pioneering work of two Victorian English scientists, Charles Darwin (1809–1882) and Alfred Russel Wallace (1823–1913), who independently discovered the principles underlying evolution (see Figure 1.2). Darwin was a wealthy English gentleman, whilst Wallace always struggled to make ends meet. Socially, they were worlds apart, but they shared a passion for observing wildlife and collecting insects, particularly beetles.

The theory of evolution introduced the ideas of competition, struggle, uncertainty, and random chance into the ordered Victorian world view. This was deeply unsettling then, and remains so now. Many other people before Darwin and Wallace had thought that new species might evolve from existing species, including Jean-Baptiste Lamarck, Robert Chambers, and Erasmus Darwin (Charles' grandfather). The difference was that none of them proposed a mechanism, backed up by evidence, to support their claims.

Charles Darwin's '*On the Origin of Species by Means of Natural Selection, or the Preservation of Favoured Races in the Struggle for Life*' was published in 1859. It became, and remains, a cornerstone of evolutionary thinking—partly because of the elegance of Darwin's writing that has captivated the imagination of readers for generations. Darwin speculated on the magnificence of the interconnectedness of species—because all were related through common ancestry to a possible single common ancestor:

> *"There is grandeur in this view of life, with its several powers, having been originally breathed into a few forms or into one."*

Humans were now one species competing amongst many, no longer the pinnacle of Creation. Above all, the principles of evolution describe powerful processes at work that:

> *"from so simple a beginning endless forms most beautiful and most wonderful have been, and are being, evolved."*

Figure 1.2a&b Charles Darwin and Alfred Russel Wallace were both active scientists who travelled the world in search of specimens.

(a)

(b)

Source: George Richmond, from Leakey, R., and Lewin, R. (1977). *Origins*. London: Penguin Books (1.2a) and Marchant, J. (1916) *Alfred Russel Wallace—Letters and Reminiscences, Vol. 1*. London, New York, Toronto, and Melbourne: Cassell and Company. Plate between pp. 36–37 (1.2b).

This book will consider just a few of the 'endless forms most beautiful' to tell the story of evolution from the perspective of contemporary science. Evolution has become the major unifying principle of the biological sciences. In 1973, Dobzhansky, an architect of the 'modern evolutionary synthesis', wrote that nothing in biology makes sense except in the light of evolution, and recent advances in molecular biology, especially genomics, only serve to strengthen that claim.

This book will consider the mechanisms for evolution, including the birth and death of species. The different types of evidence that support evolution will be considered, including the latest developments to our thinking about evolution.

Origin of Species did not discuss human evolution, but both Darwin and Wallace later wrote extensively about it. Recent advances in genomics, radiocarbon dating of fossils, and cultural anthropology are all making significant impacts on our understanding of our own story, the continuing evolution of the human species. We will consider these advances in the final chapters of this book.

The 'gap year' that changed the world

In December of 1831, after he graduated from University, Charles Darwin set out on a sort of gap year experience which leaves modern jet-setters in the shade. With only an ordinary degree and a geology summer course behind him (he was too busy eating, drinking, hunting, and collecting beetles to spend much time on his studies), he was the 'self-funded gentleman naturalist' chosen by Captain Robert Fitzroy to accompany him on a journey to South America on the surveying ship HMS Beagle (see Figure 1.3).

Fitzroy himself was only 26 years old, in charge of 30 men, and he wanted someone who was not a direct member of the crew to keep him company throughout the voyage. Darwin was offered the chance to record the natural history and geology of the places they visited, while others on the HMS Beagle did hydrographical surveys and mapped coastlines for the British Navy. The two-year round-the-world voyage commenced in 1831, and stretched out to five years. Despite being seasick much of the time, Darwin kept meticulous journals, and regularly sent copies, together with preserved specimens, back to England.

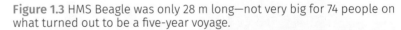

Figure 1.3 HMS Beagle was only 28 m long—not very big for 74 people on what turned out to be a five-year voyage.

Evolution was not Darwin's idea: it was already being discussed before he was born. His grandfather Erasmus had even written in his book *Zoonomia*, 'would it be too bold to imagine, that in the great length of time, since the earth began to exist, perhaps millions of years ... that all warm-blooded animals have arisen from one living filament'. The Victorian era in Britain was a time of tremendous socioeconomic change, with the Industrial Revolution in full swing. Agriculture was changing out of recognition from what it had been like for millennia. Darwin had grown up watching farmers improve their livestock by selective breeding, and oil paintings of prize livestock would be displayed amongst family portraits (see Figure 1.4).

Earth-shattering insights

Charles Darwin's experiences on his gap year changed his view of the world. He experienced an earthquake in Chile in 1835, and saw signs that the land had just been raised, including mussel-beds stranded above high tide. High in the Andes he saw seashell fossils. On rides into the South American interior, he realized that the strange stepped plains were in fact ancient raised beaches of shingle and seashells. He became aware of the vast periods of time that had passed, and of the cataclysmic changes experienced by the land in that time.

Reaching the Galapagos Islands off the northwest coast of South America he, like other visitors, was amazed that locals were able to identify which

Figure 1.4 Wealthy British landowners competed to breed the largest animals, and it was seen as a patriotic act to feed the growing population.

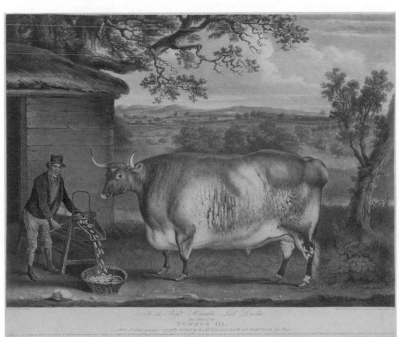

Source: Yale Center for British Art, Paul Mellon Collection

island a giant tortoise came from simply by its appearance! But by now Darwin's imagination had been captured by earthquakes and geological changes. He was intrigued by the development of landforms—and the Galapagos are volcanic islands. He and his assistant hurriedly collected and preserved specimens so they could be classified later. The 'Darwin's Finch' story wouldn't start until long after he returned home.

What is a species?

In evolution, we constantly refer to different species—but how is a species defined and recognized? Even now, every new species needs to be described in a published paper. The unveiling of a new species begins with the formal description of a 'type specimen' which will then be stored somewhere such as the research collection of a major museum. The author of the article will seek to place it in the classification hierarchy and will give it a unique taxonomic name. Once the paper is accepted for publication, other scientists can check that it is indeed a new species.

On Darwin's return, he gave the preserved bird specimens he had made on his travels (see Figure 1.5) to ornithologist John Gould to be identified and described. At the next meeting of the Zoological Society of London, Gould reported that the Galapagos birds that Darwin had thought were grosbeaks, blackbirds, and finches were actually 'a series of ground Finches which are so peculiar [as to form] an entirely new group, containing 12 species'. The story made the newspapers.

Figure 1.5 These scruffy stuffed birds are some of the actual specimens Charles Darwin prepared on the Galapagos all those years ago.

Source: Chronicle/Alamy Stock Photo

Today we call this group of finches 'Darwin's Finches', although Darwin himself had no idea of their future significance when he collected them. In fact, it was Gould who suggested they may have all descended from finches which had been blown there long ago. Remembering the unique Galapagos tortoises and how each island had its own distinct variety, Darwin wondered whether the same might be true of the finches. Unfortunately, because he had not realized this at the time, his Galapagos specimens had become mixed up, and he had to compare his bird collection with those of other members of the expedition in order to figure out which birds had come from which islands! Eventually the model of the evolution of Darwin's finches by natural selection, which we are so familiar with today, emerged from the chaos of scruffy specimens, muddled notes and conversations with other biologists (see Figure 1.6).

Figure 1.6 The Galapagos Islands off the coast of Ecuador in northwest South America (a) were originally colonized by a small brown ground finch from the mainland. The different conditions on the islands favoured different traits, and one species evolved into twelve, including the large ground finch (b) and the warbler finch (c) shown here.

(a)

Source: b) Manakin/iStock and c) Kongsak Sumano/Shutterstock

Case study 1.1
The Galapagos story ... continued!

Darwin's finches have captured people's imaginations ever since he first described them. In the 1970s they hooked Peter and Rosemary Grant, evolutionary biologists who were interested in the forces that drive evolution. They
thought Daphne Major, one of the smallest islands of the Galapagos, would
be a perfect place to study selective forces because it is uninhabited, and so
small they would be able to become familiar with every single bird on the
island. Like Darwin, they went for two years: their research there lasted 40!

They began their studies on Daphne Major in 1973. Every bird was captured,
a numbered ring placed on its leg, and they recorded other features such
as beak size. The birds were called medium ground finches, *Geospiza fortis*.
Their diet was mainly seeds, and they showed considerable variation of beak
sizes, from relatively small to relatively large. In 1977, there was a prolonged
drought and many of the plants on the island died. Those that were left had
very large, hard seeds. Heartbroken, the Grants could only watch as natural
selection took place before their eyes and dozens of birds died of starvation.
By the end of the drought they were amazed to see that the distribution of
beak sizes in those medium ground finch adults which had bred the previous
season had changed completely.

Look at Figure A. Can you see what happened during the drought? Only
the birds with the bigger beak sizes survived. Furthermore, the distribution
of beak sizes in the offspring was completely different to what it had been
before the drought: the modal group had moved from 8.8mm to 9.8mm! Why
do you think this happened?

In 2003, a drought similar in severity to the 1977 drought occurred on the
island. However, late in 1982 a breeding population of large ground finches
(*Geospiza magnirostris*) had become established on the island. This species
has a diet overlap with the medium ground finch (*G. fortis*) and were competing for the larger seeds. Following the drought, the medium ground finch
population had a decline in average beak size, in contrast to the increase in
size found following the 1977 drought (see Figure B).

Competition leads to character displacement

The Grants hypothesized that the smaller-beaked individuals of the medium
ground finch may have been able to survive better this time by being able
to eat smaller seeds, avoiding competition for large seeds with the larger
ground finches *G. magnirostris*. The 2003 drought and resulting decrease in
seed supply may have increased competition between *G. fortis* and *G. magnirostris*, particularly for the larger seeds which had enabled survival of the
medium ground finches 25 years earlier. As they had larger beaks, the population of *G. magnirostris* had a competitive advantage when it came to eating
the larger seeds. This led to the *decrease* in average beak size among *G. fortis*
despite the drought conditions being very similar.

Darwin had hypothesized that the changes leading to speciation happen very
slowly. Looking at the wildlife around us we get the impression that species stay

Figure A Distribution of beak sizes of medium ground finches on Daphne Major.

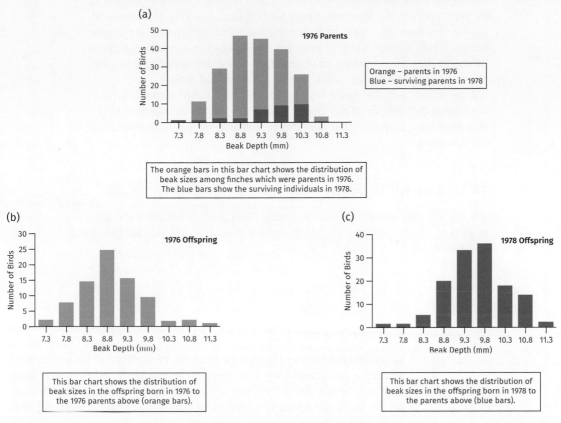

(a)

1976 Parents

Orange – parents in 1976
Blue – surviving parents in 1978

The orange bars in this bar chart shows the distribution of beak sizes among finches which were parents in 1976. The blue bars show the surviving individuals in 1978.

(b)

1976 Offspring

This bar chart shows the distribution of beak sizes in the offspring born in 1976 to the 1976 parents above (orange bars).

(c)

1978 Offspring

This bar chart shows the distribution of beak sizes in the offspring born in 1978 to the parents above (blue bars).

Source: Grant, P.R., and Grant, B.R.. (2003). What Darwin's Finches can teach us about the evolutionary origin and regulation of biodiversity. *BioScience* 53: 965–975.

Figure B Graph to show distribution of beak size of medium ground finches (*G. fortis*) over time on Daphne Major.

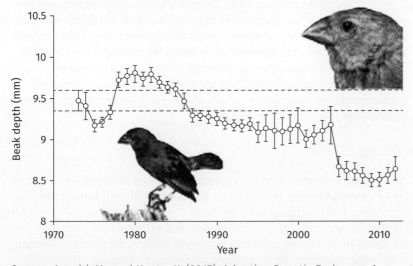

Source: Arnold, M., and Kunte, K. (2017). Adaptive Genetic Exchange: A Tangled History of Admixture and Evolutionary Innovation. *Trends in Ecology & Evolution*. 32.

the same from year to year, and this is probably why most people in the time of Darwin, including Darwin himself until after he'd returned from the Galapagos, did not believe in the **transmutation** (changing from one kind to another) of species. However, the Grants have demonstrated that perhaps we are simply not observant enough. Close observation of a species in the wild can show significant fluctuation in appearance, caused by the selection pressure of a mix of biotic and abiotic factors, over a relatively short period of time.

As Jonathan Losos, a Harvard evolutionary biologist, has observed, 'Perhaps the biggest contribution of the Grants' work is simply the realization not only that evolution can be studied in real-time, but that evolution doesn't read the textbooks.'

Pause for thought

The average PhD research project lasts only three years. What are the implications of this for studying long-term processes such as ecological change and evolution?

Darwin's world

To understand Charles Darwin, we must try to understand his world. Both his father and grandfather were doctors, and so birth, sickness, and death were very much part of the world he grew up in. During their lifetimes, Britain had experienced considerable demographic growth, doubling from 7.1 to 14.2 million people during the century after 1721.

Exciting tales of limitless fertile land were filtering into Europe from the Americas. The American Benjamin Franklin had predicted exponential population growth as early as 1751, as he realized there were not the same limits to population growth in America as there were in Britain.

In 1798, the mathematician vicar, Thomas Malthus, published his first version of *An Essay on the Principle of Population*. Whereas others were looking to expansionism, he considered what happens when there is not enough land to support everyone. Inevitably this would lead to high infant mortality, shortened adult lifespan, disease, and even to famine and starvation.

Malthus was worried: Britain had lost the American colonies and yet the British population continued to climb. The Hungry Forties and Irish Potato Famine were just around the corner. . . .

Building a case

When Charles returned from his voyage, his wealthy relations set up a trust fund to provide him an income so he could throw himself into sorting out, identifying, and describing the thousands of specimens he had collected. All the time he was trying to understand how the extraordinary variety of living things he had found came into existence, at the same time as reflecting, in a number of articles about geology, on the scale of geological time.

While aboard the Beagle he had thought that the diversity of organisms arose from there being various 'loci of creation'. Within a year, however he was beginning to accept the theory of transmutation (see Figure 1.7).

Less than two years after returning he read Malthus' essay 'for amusement', and later recorded in his autobiography, 'It struck me that under these circumstances, favourable variations would tend to be preserved and unfavourable ones to be destroyed. The result of this would be the formation of new species. ... Here, then, I had at last got a theory by which to work.' Malthus' essay was about human society and economics: Darwin had realized that overpopulation and competition were not just found in people but were actually part of a universal law which applied to all living things.

In 1839, Charles had a momentous year. He married his cousin Emma and published what would later become known as *The Voyage of the Beagle*. He could see in his mind's eye that the variety of life he had experienced must have arisen over vast amounts of time. Where to start collecting evidence about the *process* of change? To test a theory, you need evidence. He turned to the world in which he lived, and focused on how domestic animals are selected to form new varieties. He published a pamphlet called '*Questions about the Breeding of Animals*', gave it out to friends and acquaintances who bred various species of domestic animals, and waited for the responses to arrive. Realizing that data was what he needed, he pleads at the end of the first paragraph, 'Always please to give as many examples as possible ... to illustrate these ... questions.'

Bear in mind that Mendel, thirteen years younger than Darwin, would not start his work crossing peas until 1854. So although Darwin knew about 'characteristics', the idea of genes had not yet been discovered. Most people thought that characters from both parents would be blended in their progeny, and this blending and dilution would continue each generation.

Only three years later Darwin found himself in the middle of a global experiment, caused by unusual weather which led to the failure of potato, wheat, rye, and oat harvests across Europe over several years. The Irish Potato Famine was part of this. About a million people died in two years. Relatively few people died of actual starvation as a result of the rotted potatoes. Most succumbed to hunger-related diseases such as dysentery, typhoid fever, and typhus. Scotland was also affected, and this triggered the second Highland Clearings. Malthus' writings had proved to be frighteningly accurate.

During the 1840s and 1850s Darwin wrote in a wide variety of publications, from the *Gardeners' Chronicle* to the *Proceedings of the Royal Society*, the most prestigious scientific organization in the country. His articles include letters to the *Gardeners' Chronicle* requesting lizards' eggs (so he could do experiments to see if they float in seawater), and a number of items about his experiments to assess the viability of plant seeds when immersed in seawater. He also wrote an excited response to a report in the *Chronicle* that live specimens of the snail *Zua lubrica* had fallen during a thunderstorm in Winchester.

"I hope 'C.' will forgive me for suggesting to him how very desirable it is that so extraordinary and very interesting a fact should be authenticated. ... It is really almost a duty towards the science. ..."

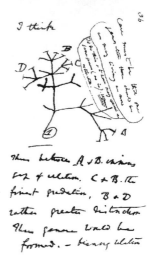

Figure 1.7 In his notebook B of 1837, written the year after he returned from his voyage, you can see how the young Darwin's ideas are developing. Here is his first sketch of an evolutionary tree.

For two decades Darwin painstakingly gathered evidence, trying to understand how species dispersed, how natural selection worked, and how it could lead to evolution. Perhaps the piece of evidence which finally persuaded him came from breeding pigeons. He was astonished when, only a year after acquiring his first fancy pigeon breeds, he succeeded in 1856 in breeding pigeons that looked exactly like the Rock Dove, which he reckoned was the original ancestor of all domestic pigeon breeds (see Figure 1.8).

Figure 1.8 Darwin was initially very reluctant to breed pigeons but was soon captivated.

Source: Chronicle/Alamy Stock Photo

Race to the finish

As a 25-year-old, Alfred Russel Wallace (see Figure 1.2) spent four years in the Amazon looking for evidence of transmutation of species, funding himself by selling specimens to collectors (see Figure 1.9). He lost everything when he was shipwrecked on the way home, but he later started again in the Malay Archipelago. In 1855, Wallace published an article about the 'introduction of species' in which he made guarded comments about evolution.

Darwin's mentor, geologist Charles Lyell, showed Wallace's article to Darwin and urged him to publish as soon as possible. Darwin started writing a book which he planned to call *Natural Selection*.

Figure 1.9 Part of Wallace's collection of 80,000 beetle species, which helped him develop his ideas of both evolution and warning coloration. They are now in the Natural History Museum in London.

Source: Natural History Museum: Coleoptera Section / Flickr

In early 1858, Wallace wrote *On The Tendency of Varieties to Depart Indefinitely from the Original Type* and sent it to Darwin with a request to send it on to Lyell for comments. Darwin panicked, writing in his covering letter, '*your words have come true with a vengeance, ... forestalled. ... all my originality, whatever it may amount to, will be smashed*'. Lyell counselled a strong nerve and urged Darwin to publish as soon as possible. But Charles's baby son, who had Down's syndrome, was dying of scarlet fever, and he could think of nothing else. The only proof he had that he had not borrowed his ideas from Wallace was an essay he'd written in 1844 and shown a few friends. He sent this to Lyell, and in July this paper and Wallace's were read at the meeting of the Linnaean Society. Neither Darwin nor Wallace were present, and the Linnaean Society was so unimpressed they announced no work of any exceptional merit was reported that year!

Spurred into action, Darwin rushed out a manuscript called *On the Origin of Species*. Reviewers were unhappy with it as they felt it covered too much and too superficially. Darwin put his foot down. He had spent over six months on *Origin* already and was on the last chapter. Scientific discoveries are usually announced in specialist scientific journals, and Darwin had been planning to publish a large book in a more scientific style, packed with evidence. But there was no time for polish. The result is a conversational style, accessible by non-scientists as well as scientists. The last paragraph begins:

"*It is interesting to contemplate a tangled bank, clothed with many plants of many kinds, with birds singing on the bushes, with various*

*insects flitting about, and with worms crawling through the damp
earth, and to reflect that these elaborately constructed forms, so different from each other, and dependent upon each other in so complex
a manner, have all been produced by laws acting around us."*

On the Origin of Species was published in November 1859, followed
barely two months later by a second edition. Is it any surprise that a de-
tective story, written in such beautiful prose, sold out immediately? Or that
Darwin's name is the one that is linked so closely with natural selection, the
mechanism which underpins evolution all around us?

Meanwhile, in Moravia …

… a young man called Gregor Mendel was about to change the way the
world looked at inheritance, and provide the evidence which explained
how natural selection worked. Mendel's family were farmers in the Czech
Republic, and he had gone to university to study philosophy and physics.
The same harvest failures which were causing so much hardship in Britain
and Ireland meant his family were struggling financially, and could not
afford his university fees. His lecturers had been impressed by his ability
and enthusiasm. As his health was too poor to farm, and he wanted to be a
physics teacher, they suggested he could join the Augustinian friars, who
would pay for him to complete his training. He later wrote that this spared
him the 'perpetual anxiety about a means of livelihood'.

The abbot, Cyrill Napp, was active in the regional agricultural society
and interested in improving food plants and animals. He soon put Mendel
to work improving the various crops grown to feed the monks—even having
a heated greenhouse built for Mendel's experiments, a real luxury and no
small expense in 1855. Did Napp realize even then the future significance
of Mendel's work? In a photograph of the community taken in about 1862
(Figure 1.10). Mendel has been placed directly behind Napp and is holding a
flower. In 1865, Mendel presented his results in a talk to the local scientific
society and published a paper the following year.

In it, he described 'factors of heredity' which behaved in certain math-
ematical ways when they were passed on to the offspring. This was revolutionary because the received wisdom at the time was that the parents'
characteristics were smoothly blended in the offspring, a bit like mixing
paint.

Mendel described three principles which governed the way these factors
are passed on:

1) **Law of Segregation:** Each hereditary characteristic is controlled by
   two 'factors' which separate into different gametes when they are
   formed, and randomly reunite during fertilization.

2) **Law of Independent Assortment:** During gamete formation the separation of factors for one gene does not affect factor separation for
   another gene.

3) **Principle of Dominance:** If a pair of factors are different, the
   dominant form will mask the effect of the recessive form.

Figure 1.10 In this posed photograph, Mendel is standing behind the head of his monastery, holding a flower. Interestingly, the attention of the other monks around Mendel is clearly focused on him.

Source: Courtesy of the Mendel Museum of Masaryk University, Brno, Czech Republic

It seems that the scientific world was not ready for Mendel's work, despite the fact that his paper was published and widely circulated. Mendel continued his experiments for a while, but many traits such as those which show continuous variation do not follow the simple arithmetic ratios discovered by Mendel. The significance of his work was not widely recognized for over 30 years—long after his death in 1884. Eventually scientists working to understand the new theory of discontinuous inheritance rediscovered his paper and—in a very gracious act—ensured that it was Mendel who got the credit rather than themselves.

Intriguingly, there is a German copy of Darwin's *On the Origin of Species* in the Moravian Museum in Brno that belonged to Mendel. In places he has underlined sections. It is tempting to wonder what Mendel would have said to Darwin had they met!

Selection: the engine which drives evolution

When the Grants were measuring the changing beak sizes of finches in the years following the 1977 drought (see Case study 1.1), they captured a moment in time when directional selection was taking place. In that instance, the selection pressure was very extreme: a large proportion of the population on that island died, and with them a portion of the gene pool for smaller beaks. The survivors bred the following year: although the spread of beak sizes remained the same the beak size of the modal group had increased. So what is directional selection?

Directional selection

Directional selection is probably what most people think of when they think of natural selection. Darwin had shown it with his pigeon breeding experiments. In the years following publication of *Origin*, there were scientists everywhere looking for evidence which would catch evolution in action. Here are some examples.

The peppered moth: Darwin's theory solves a mystery

The Victorians loved collecting things. Wallace earned a living by selling exotic insects. Maybe some were already wondering why the melanic (dark) forms of butterflies and moths such as the peppered moth (*Biston betularia*—see Figure 1.11) had been becoming more than just an occasional rarity. The peppered moth spends much of its time resting on lichen-covered tree trunks, camouflaged from the birds that would otherwise eat it.

The first *carbonaria* (melanic) morph of *B. betularia* was recorded in Manchester in 1848, and by 1895 it comprised 95% of peppered moths there. In some industrial parts of England, the original pale wild type was almost lost by the end of the century.

Darwin's theory gave scientists the foundation for a hypothesis. By 1877, the Scottish entomologist Francis Buchanan White was suggesting that the dark forms of the peppered moth were protected in some way. By 1894, James Tutt had suggested that industrialization, with its attendant soot and pollution, had created a changing environment in which the melanic form now had a selective advantage. Both buildings and tree trunks became blackened. Lichen is extremely sensitive to pollution and so disappeared from the tree trunks on which *Biston betularia* rested. Predators could now clearly see the pale form of the peppered moth, but the darker moths were camouflaged (see Figure 1.12).

Figure 1.11 The pale form of *B. betularia* is almost invisible on the pale lichen. Imagine how it would show up against the dark building in Fig 1.12 - and how camouflaged the dark form becomes.

Figure 1.12 Britain's urban buildings were blackened with soot from coal fires during the 19th and early 20th centuries. In the late 20th century contractors were paid to remove the sooty coverings. The difference is clear!

Source: Visual Resources Centre at Manchester Metropolitan University

The discovery of genes: the missing piece of the puzzle

Any investigation of selective processes was handicapped by the fact that no one knew about genes. Mendel's article, written in German, was mouldering in various libraries. Darwin's own theory of inheritance was pangenesis: he thought that organic particles which he named gemmules were shed by all the organs of the body and carried in the bloodstream to the reproductive organs where they accumulated in the gametes. When passed on to offspring, these gemmules would develop into each part of an organism. Each individual would therefore have a unique mixture of its parents' gemmules, and therefore would express characteristics inherited from both parents.

The rediscovery of Mendel's work in 1900 not only introduced the idea that the particles of inheritance were discrete and constant from generation to generation, it also introduced the idea of dominant and recessive characters. This was a game-changer, but nevertheless it was not accepted by everybody.

In Britain, William Bateson became a leading champion of Mendel's theories and gathered an enthusiastic group of followers. Known as 'Mendelians', its supporters initially clashed with Darwinians (supporters of Charles Darwin's theories which included his concept of gemmules). The Mendelians realized they needed to discover whether characteristics were inherited in a blended form (as they would be with gemmules) or a discontinuous form as Mendel had shown in peas. Bateson coined the term 'genetics' but the term did not really catch on until the botanist Wilhelm Johannsen suggested that they call Mendelian factors of heredity 'genes'. He also made the distinction between the genetic traits of an individual (genotype) and its outward appearance (phenotype).

By 1914, William Bowater realized that melanism in peppered moths was inherited through a single gene which had a number of forms or alleles, which included pale (*typica*) through intermediate (*insularia*) to dark (*carbonaria*). Predation experiments, particularly by Bernard Kettlewell in the early 1950s, established that in an unpolluted Dorset woodland, birds preyed more on the *carbonaria* morph, but in a polluted Birmingham woodland more preyed on the *typica* morph. His study concluded that 'industrial melanism in moths is the most striking evolutionary phenomenon ever actually witnessed in any organism, animal or plant'.

Soon an inadvertent national experiment took place: the 1956 Clean Air Act was passed. Sure enough, the black form of the moth began to decline and the pale form, which was better camouflaged on lichen-covered tree bark, became more numerous. In only 50 years the frequency of the *carbonaria* form decreased in the Manchester area from over 95% to less than 5% (see Figure 1.13).

Are birdfeeders a selection pressure?

The great tits in Oxford's Wytham Woods may be the world's most intensively studied wild birds. Oxford University has owned the woodland since 1942, and being on their doorstep has simplified fieldwork there. Similar long-term work in the Netherlands has enabled researchers to

Figure 1.13 As soot and sulphur dioxide disappeared from the atmosphere lichens were able to grow again, resulting in the resurgence of the *typica* form and a fall in *carbonaria*, shown here.

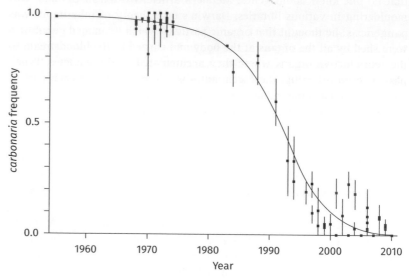

Source: Roger L H Dennis

compare populations of great tits (*Parus major*, see Figure 1.14) in the two countries. They were surprised to find that since 1970 the beaks on the British birds have become longer than those of the Dutch birds. This has been linked to changes in genes governing both beak length and face shape. Furthermore, the research team discovered that birds which have the gene variants for longer beaks visit birdfeeders more than those without the variants.

Figure 1.14 Great tits are one of the most common birds visiting garden bird-feeders, according to the annual RSPB birdwatch census. © Anthony Short

Cepaea—the snail that looks like a boiled sweet selection

Cepaea nemoralis (see Figure 1.15) is a snail which comes in a variety of shapes and colours—put more scientifically, it is polymorphic for a number of genes affecting colour (yellow/pink/brown), number of bands (none to five), body colour (pale/dark), and others. In the early heady days of ecological genetics, scientists exploring the newly acquired Wytham Woods and surrounding areas noticed that the snails' appearances were more variable in grassland than on the uniformly brown background of dead woodland leaves. What patterns can you see in Figure 1.15?

Were the snails cryptically coloured? Were the avian predators acting as a selective force to create an unbanded pink/brown population in woodlands

Figure 1.15 Judged on their colour alone, these *Cepaea nemoralis* snails might appear to be several species—but they are all examples of the same polymorphic species. By scoring the colour and habitat of the snails, Cain and Sheppard found patterns in the distribution of different *Cepaea* morphs.

(a)

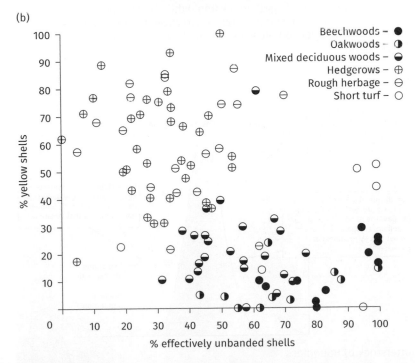

(b)

Source: a) Eric Isselee/Shutterstock b) A.J Cain & P.M Sheppard

but a yellow/banded population in hedgerows and fields? The percentage of banded shell fragments by a thrush anvil in a woodland was significantly higher than the percentage of banded snails in the population as a whole. Releasing a mix of colours in Wytham Wood and then collecting shells from thrush 'anvils' seemed to indicate that thrushes chose yellow snails, being more visible against the dead leaves of the forest floor, over the darker ones.

Other patterns were discovered. For instance, yellow shells are more common in warmer climates (the pale yellow reflects sunlight) but shells that are darker or banded are more common in colder climates. This will help the snails warm up enough to be active, as they would absorb more heat from the sunlight.

Stabilizing selection

One of the remarkable things which Darwin observed was the variety hidden in most populations. So how is it hiding? One way is stabilizing selection, which is directional selection against both the extreme phenotypes so intermediate variants have the highest fitness. The clutch size (number of eggs) of birds show a lot of variation between species, but is remarkably consistent within a species (see Figure 1.16).

Diversity in a population is decreased due to stabilizing selection, which will reduce and can eliminate the genotypes at the extremes. However, mutations add to the genetic diversity in a population. This makes it more likely that if the environment changes, some individuals will have favoured phenotypes and survive, and over successive generations the population will become better adapted to the new environment.

Figure 1.16 Many species of birds lay a similar number of eggs—for example this song thrush nest has a typical clutch of 3 to 5 eggs. Stabilizing selection is the result of different directional selection pressures operating in different directions. In this case the selection pressures include competition between the nestlings for food and space, and predation.

(b)

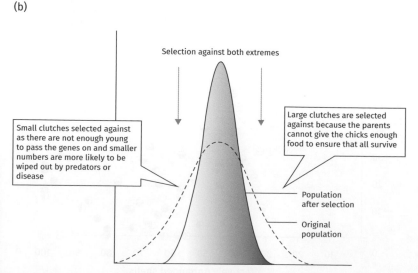

Source: a) Vishnevskiy Vasily/Shutterstock b) Boundless

The Goldilocks effect and goldenrod gall flies

Goldenrod gall flies (*Eurosta solidaginis*) lay their eggs in the young shoots of the goldenrod plant (*Solidago* sp.). When the larva hatches, it makes a chemical that causes the plant tissue to swell around it, forming a gall. The larva lives inside the gall, eating the soft plant tissue and eventually pupating. The galls keep the larvae safe from everything except predatory birds and parasitic wasps. Predatory birds are more likely to attack large galls. Parasitic wasps such as *Eurotema* sp inject their own egg into the gall fly egg or larva, and only the walls of small galls are thin enough for the ovipositors of parasitic wasps to penetrate. Only galls of intermediate size remain intact because they are less attractive to both predators and parasitic wasps. So, selection pressures act on the extreme gall sizes and only larvae in galls of intermediate size survive (Figure 1.17).

Characteristics like these are caused by a number of genes acting together (polygenic), to create the phenotype upon which stabilizing selection acts. Over time, some of the genes that control the characteristic can be turned off or masked by other genes (eg epistasis). In this way selection is fine-tuned to enable subtle variations so that the population can quickly adjust to new or different environmental variables.

An example is coat colour in the house mouse, which is determined by several genes acting together. The gene which determines the background colour (black/brown) is epistatic to a colour intensity gene which can modify the coat to be pale brown or grey. How might these genes work together to enable survival in a variety of habitats?

Figure 1.17 The stabilizing selection affecting the amount of gall-inducing chemical secreted by gallfly larvae is the result of opposing pressures from two different predators. Copyright © 2016 The Open University

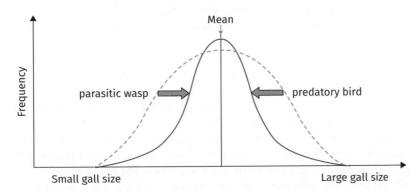

Disruptive selection

Disruptive selection is the opposite of stabilizing selection, because the intermediate phenotypes are selected against.

Darwin's finches: each island is an experiment

The finches on islands in the Galapagos which had plants with both large or small seeds will have experienced disruptive selection. Finches with small beaks would find it easier to pick up the small seeds while those with large beaks would find it easier to crack open large seeds. Birds with intermediate-sized beaks were at a disadvantage. In the end it was disruptive selection which drove speciation to make the specialists—large-beaked finches which feed on large, tough seeds and small-beaked finches which eat the smaller seeds. Continued selection against intermediate beak size then keeps the selection pressures up to continue to separate the newly formed species.

The strength of those selective forces is not always the same, however. In years when there was an abundance of small seeds, there would be more than enough to go around the whole population, even those with larger beaks. During these years there might be no disruptive selection pressures operating on beak size. But the food availability will fluctuate year to year, and so specialization in food resources will have proceeded in fits and starts.

One of the wonderful aspects of the Galapagos finches for scientists is that the different islands are like parallel experiments. You can see in Figure 1.18 how this works, by comparing the beak depths of the small and large ground finches (*G. fuliginosa* and *G. fortis*).

Lazuli buntings: don't look too sexy!

The striking lazuli bunting, *Passerina amoena* (Figure 1.19a), lives in the western half of North America. The plumage of males becomes a vivid blue as they mature: those with the brightest plumage are generally more successful at winning the highest quality territories.

The feather brightness of yearling males varies, ranging from brown to patchy blue. When there are limited high-quality nesting sites, the adult males will tolerate the dull yearlings establishing breeding territories nearby. This means that both brightest males and the dullest yearling males can succeed in establishing territories, and therefore attract females. Those yearling males which have an intermediate plumage are unable to obtain territories and mate (see Figure 1.19 b).

Paternity analysis of the lazuli bunting nestlings using DNA fingerprinting demonstrated that 49% of the nests contained chicks with more than one father, so males have a good reason not to trust their mates. All the evidence indicates that the successful mature males tolerate the dull yearlings nesting nearby because they will not be so tempting to their mates when territories are crowded!

Figure 1.18 On Daphne and Los Hermanos the beak sizes of *G. fuliginosa* and *G. fortis* cover a similar intermediate size range, but on Floreana and San Cristobal they are sympatric and so hardly any of either species have intermediate beak sizes.

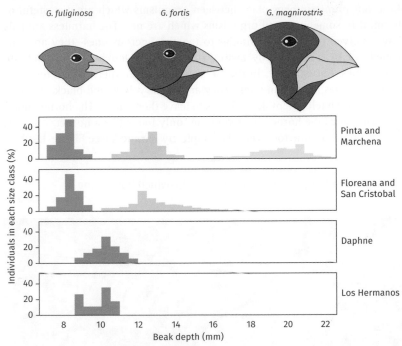

Source: Lack, D. (1947). *Darwin's Finches*. Cambridge: Cambridge University Press.

Figure 1.19 Lazuli buntings are so named because the head of the male is the colour of the gemstone lapis lazuli. The female is dull-brown. The brightest coloured, successful males leave young, dull brown, males alone because they are judged to be less of a threat.

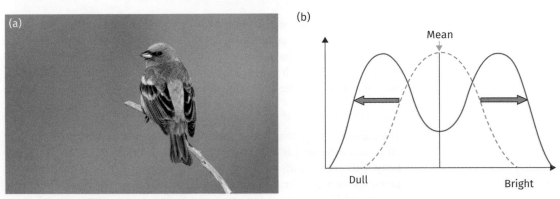

Source: a) Agnieszka Bacal/Shutterstock

Batesian mimicry—sheep in wolves' clothing!

Sometimes the striking colour patterns of organisms are due to Batesian mimicry. This form of mimicry was first described by Henry Bates, who did a lot of work in South America with Alfred Russel Wallace. Batesian mimicry takes place between organisms which are distasteful or harmful in some way, and organisms which are not. The harmless animals have evolved to look very similar to more dangerous ones–they are protected not by a sting or by tasting nasty, but by looking like something which does. For example, in the UK you may see hornets, which look like very large wasps and can sting. They are bright yellow and black, warning colours which tell potential predators to leave them alone. The hornet moth, also known as the hornet clearwing, not only looks like a hornet, it moves and behaves like one too—yet it is completely harmless (see Figure 1.20).

Figure 1.20 Batesian mimicry between the hornet and the hornet moth—which has a sting? © Anthony Short

Sexual selection

Disruptive selection was combined with sexual selection in the lazuli buntings. Birds are a popular subject for studying sexual selection because so many bird species have outrageous male plumage. In sexual selection, one sex has features which increase its attractiveness to the opposite sex. Often, but not always, it is the males who compete and the females who choose.

The widowbird: it's the tail that counts

The tail of the male long-tailed widowbird (*Euplectes progne*) seems to defy all common sense: his tail makes up about two thirds of the length of his body. This seems quite a handicap if you are considering vulnerability to predators as a significant factor in natural selection. The female's streaky tawny and black body is far better camouflaged and seems a much more sensible choice. So does the long tail add to his reproductive success? In 1982, Malte Andersson gave some lucky males tail extensions by snipping the tail feathers of some captured males in half and added the feathers to the tails of others. He compared the reproductive success of birds with normal tails with birds whose tails had been artificially shortened or lengthened. You can see the results in Figure 1.21.

By the end of the season the male widowbirds with tail extensions had doubled the number of nests with mates whereas success of the ones with shortened tails had halved.

Figure 1.21 If males with extra-long tails are able to collect more mates, what do you think the opposing selective forces on tail length might be?

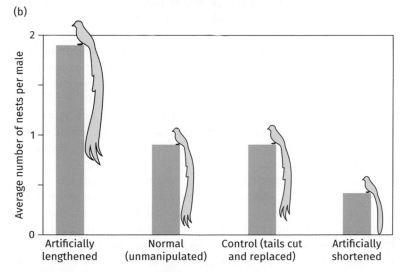

Source: a) Bernard Dupont/Wikimedia b) Hillis et al. (2014). *Principles of Life*. New York: WH Freeman.

Cichlid fish: dazzling her with colour

The aquaria of the world would be very dull if it were not for cichlid fish (see Figure 1.22).

There are over 500 species of cichlids in East Africa's Lake Victoria alone. This diversity is partly due to the extraordinary geological history of that

Figure 1.22 With so many competing species, male cichlids have evolved an absurd variety of coloration.

Source: Arunee Rodloy/Shutterstock

area, but the bright colours and varied patterns of male cichlids have been developed and refined by fierce sexual selection. The value of sexual selection in maintaining the bright colours has become sadly all too clear now that the waters of Lake Victoria are becoming murkier due to erosion and pollution.

Recent research shows that females of *Pundamilia nyererei* from clear areas of the lake strongly prefer to mate with brightly coloured red males whereas females from areas with cloudy water were more likely to accept dull-coloured males as mates. In time some species distinctions may disappear without this pre-zygotic sexual selection. Indeed, even now *P. nyererei* females in the muddiest areas sometimes mate and hybridize with a closely related species (*Pundamilia pundamilia*) which has blue males instead of the red of their own species.

Every species is experiencing selective pressures shaping its physiology, anatomy, and behaviour almost every moment. Often there are multiple selection pressures operating simultaneously, and they may change in the short term or in the long term. It is like a rather complicated dance with its biotic and abiotic environment in which there are many moves, but the dance itself is changing. As a result, each species may be a picture, but the picture is constantly changing—not just over millennia but even during a year. Over evolutionary time we can look back and see the big picture changing too, as natural selection leads not only to speciation but also to extinction.

Chapter Summary

- In 1831, Charles Darwin joined a five-year round-the-world expedition during which he collected geological and biological evidence which convinced him of transmutation of species.
- Peter and Rosemary Grant discovered evidence of Galapagos finch phenotypes changing in response to directional selective forces.
- Darwin spent about 25 years after the Beagle expedition collecting evidence to support his theory of evolution by natural selection.
- Gregor Mendel's discovery of genes came after Darwin had published *Origin of Species*, and was ignored by the scientific community until over 30 years later in 1900.
- The discovery of melanism in *Biston betularia* helped convince scientists that Darwin's theory provided workable hypotheses.
- *Cepaea nemoralis* showed the value to species of polymorphisms, as they enable the species to live in a greater variety of habitats.
- Opposing selective forces create stabilizing selection.
- In disruptive selection a further force selects against intermediate phenotypes.
- Sexual selection occurs within a species when one gender, usually females, show preference for a particular trait.

Further Reading

Desmond, A., and Moore, J. (1992). *Darwin*. London: Penguin Books. *A very readable, illustrated biography of Darwin*

https://www.nhm.ac.uk/discover/who-was-alfred-russel-wallace.html *The Natural History Museum's story of Wallace*

http://darwin-online.org.uk/biography.html *A very comprehensive website which includes facsimile copies of Darwin's manuscripts.*

Discussion Questions

1.1 The scientific process can sometimes be messy. While you need to accurately record what you do and observe, you also need to be open to surprises. Even when we feel we are being open-minded we can too easily only observe what we expect to see. Can you find examples of when Darwin was (a) open to the unexpected, and (b) not open to the unexpected?

1.2 'When histories are written they summarize the past in terms of our present understanding.' How is the story of the theory of evolution an example of this?

1.3 'Selection operates at the level of phenotype rather than gene.' What does this mean, and how do the examples in this chapter illustrate this?

2 THE BIRTH AND DEATH OF SPECIES

Throughout the history of life on Earth, scientists estimate that around 5 billion species of organisms have evolved—and more than 99% of them have disappeared again. We will never see the terrifying sight of *Tyrannosaurus rex* striding towards us (see Figure 2.1), or the towering beauty of the horsetail and tree fern forests which gave us coal. Our eyes will never view the myriad of invertebrates—from dragonfly-like insects with wingspans of almost 1 metre, to the tiniest flies and beetles which we only know from their remains trapped in amber. The ranks of organisms which have gone before our modern species of bacteria, plants, and animals are unimaginable.

And what of today? No-one knows exactly how many species are alive on Earth now, in the 21st century. As you saw in Chapter 1, scientists estimate that we have only discovered and identified about 14% of the biodiversity of the Earth—about 1.2 million species out of around 8.1 million which we have documented and named. We also know that species are dying out at an unprecedented rate—extinctions are happening faster than discoveries.

Species come into being and go extinct all the time—that is the nature of evolution. In this chapter we will consider the forces that drive the birth of new species—and some of the factors which can drive a species to its death.

Figure 2.1 Life before the evolution of human beings—how much do we really understand of the processes which resulted in bacteria, dinosaurs—and ourselves?

Source: tonart24/Shutterstock

Reproductive isolation and speciation

Reproductive isolation is required for speciation. In Chapter 1, we looked at the different types of selective forces which act on organisms and drive evolution. But a selective force in an area will only result in a change if the population is isolated from other populations of that species. If organisms are not isolated in some way, then outbreeding with surrounding populations will even out any changes in gene frequencies, and there will be no permanent change—although there may be fluctuations such as the Grants saw in ground finch beak sizes (see Chapter 1). Permanent, long-term changes are referred to as speciation and require genetic isolation.

How does a population become isolated? Broadly speaking, scientists divide speciation into allopatric speciation and sympatric speciation.

Allopatric speciation

In allopatric speciation, genetic isolation is the result of a physical or geographical barrier to gene flow. The Galapagos Island finches diversified into species due in part to the fact that there is almost no gene flow between islands. However, much of the sequence of events of the evolution of the Darwinian finches can only be inferred. In other species, the evidence is very clear-cut.

Once populations of a species become geographically isolated, there are many different ways in which they can become so different that they form separate species. For example, they may adapt to different conditions, develop different reproductive behaviours and timings or courtship rituals, eat different food organisms, synchronize to different seasons, flower at different times or adapt for different pollinators.

Island-hopping fruitflies

The fruitflies (*Drosophila* spp) of the Hawaiian islands provide a beautiful example of how speciation has occurred over time, and demonstrate the use of new technologies to help us understand old problems. Analysing the DNA of species, especially when we have a clear measure of the frequency of mutation in a population, allows us to build a molecular clock which gives us the time sequence of speciation (see Figure 2.2).

Mitochondrial DNA is passed through the generations from mother to offspring through mitochondria in the mother's eggs. The longer two species have been separated, the more differences there will be in their base sequences. It was this mitochondrial DNA which scientists sequenced to build up the timeline of fruitfly evolution across the Hawaiian islands.

The Hawaiian islands are volcanic, produced by a hotspot under the Pacific Ocean which, as a result of continental drift, has created a string of islands, with the youngest in the east and the oldest in the west. Hawaii is the youngest island, and it still has active volcanoes. Amazingly, a phylogenetic tree based on the degree of DNA similarity in this clade showed the same sequence of DNA diversions as the geological ages of the islands that the species came from. DNA and geological evidence together indicate that an ancestral *Drosophila* was blown onto one of the oldest islands and then island-hopped eastwards as new islands became habitable, forming new species along the way.

Figure 2.2 DNA and geological evidence come together to demonstrate how *Drosophila* island-hopped eastwards with the formation of new islands.

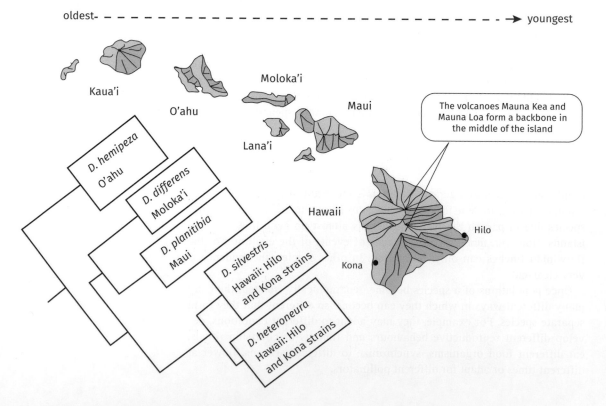

Pesky '*molestus*' mosquitoes

The barrier to gene flow doesn't have to be miles of ocean. There is a population of mosquitoes in the London Underground. They were the scourge of the 180,000 people who sheltered there from the falling bombs of World War II (1939–1945). Until quite recently, it was assumed they were the same as the mosquitoes living on the surface; however, scientists have now discovered that the two populations differ in quite fundamental ways.

The females of the surface-dwelling species, *Culex pipiens*, get their blood meals from birds. They gather in large swarms to breed, and then diapause (become dormant) during the winter. The underground mosquitoes are a subspecies, appropriately named *Culex pipiens molestus*. They do not diapause or swarm (the adults mate in pairs), and the females feed from mammalian blood rather than bird blood. Who needs a dormant period in the relatively constant warmth of the deep underground tunnels? The *molestus* females even get enough raw materials for their first brood of eggs from their larval diet, only drinking blood (often human blood), for later broods. In contrast, the surface-dwelling mosquitos need a blood meal before any eggs are laid (see Figure 2.3).

In many countries mosquitoes spread diseases between people—malaria, dengue fever, and yellow fever are just three examples. At the moment, neither of our British subspecies spread diseases—but if, as the climate changes, we begin to get mosquito-borne diseases, *molestus* mosquitoes would be a particular risk. Because they do not diapause, they would spread disease all year round.

Origins and evolution

The *molestus* mosquito is now found in underground locations all over the world. So maybe it did not evolve in London? Could it have been an underground species that moved in from elsewhere? Byrne and Nicholls used gel electrophoresis to measure variation at 20 genetic loci.

They found that the alleles in the underground mosquitoes were the same as the surface mosquitoes, implying that they were closely related to each other.

Figure 2.3 Female mosquitoes need blood meals in order to lay eggs. Deep in the underground, *Culex pipiens molestus* feeds on us!

They did not contain any of a number of alleles found in *molestus* populations in other countries. Furthermore, they found that the London Underground populations were more homogeneous than neighbouring overground populations, indicating the presence of the founder effect.

The jury is still out on this species and the debate rages on. Some scientists believe that *molestus* is a species of *Culex* rather than a subspecies of *C. pipiens*. Others believe that *molestus* originated in a more equatorial area and has moved north, carried by humans into the subterranean (and therefore relatively warm) structures we have built. Improved biochemical tests such as those comparing microsatellite DNA seem to be complicating the picture rather than clearing up the debate.

Sympatric speciation

It is relatively simple to see how geographical isolation could lead to one species separating into two new species, if the separation is consistent and continues for a long period of time.

Rather more difficult to understand is how species diverge from an ancestor when they are not physically separated. In this case there needs to be something else acting to lead to genetic isolation.

Sometimes new species arise initially through the formation of a hybrid—the result of a mating between two different species. In animals, hybrids are usually infertile, because the species have different numbers of chromosomes and so they are unable to pair during meiosis. Mules, the infertile offspring of a horse and a donkey, are a familiar example of this. Plant hybrids, however, are quite often fertile. One mechanism for this is a chromosome mutation event, such as non-disjunction leading to polyploidy when the entire chromosome set is duplicated. This ability of plants to form fertile hybrids has been one of the major drivers in the development of many of our major food crops, including wheat. Sympatric speciation has played, and continues to play, an important role in plant speciation.

There is, however, considerable debate about the importance of sympatric speciation in animals. When animals begin to use different resources—such as different food plants or different habitats—within an ecosystem, they may begin to separate and move towards forming different species. Examples are relatively rare, but sympatric speciation has doubtless played a part in the evolution of some of the animal species we see around us today.

Case study 2.1
The amazing radiation of African cichlid fish

Cichlid fish are members of the family Cichlidae, one of the largest families in the whole of the vertebrate subphylum. At least 1650 species have been catalogued and described (see Chapter 1 Figure 1.21), and new species are being discovered every year. Many of those cichlid species introduced to new areas have thrived and become nuisances: this hardiness helps make them popular for home aquaria!

Unravelling the cichlid story began with scientific amazement at the sheer variety of species in the Great African Lakes. Most cichlid species live there, and many are **endemic**. The number of closely related species speaks of a spectacular explosion of **adaptive radiation**, as each lake was colonized and species filled every available niche after the lakes formed, millions of years ago.

These African lakes owe their existence to the exceptional geological activity of the region: it is a Y-shaped boundary between tectonic plates (see Figure A). The Great Rift Valley marks part of this boundary. It is this activity which created opportunities for allopatric speciation, as changes in the shape of the landscape over geological time separated lakes from lakes, lakes from rivers, and rivers from rivers.

Scientists all over the world are looking for evidence to help them understand the sequence of the African cichlid phylogenetic tree. Using the molecular clock, they can estimate when cichlid species in a particular lake diverged from the ancestral species, and from each other, and then try to match this timescale with dating of the local geology using fossils. Genner's team explored the possibility that the Malawi cichlids had originally radiated from immigrants from the Great Ruaha river by looking for DNA relationships between the Ruaha cichlid *Astatotilapia* and the Malawi cichlids.

Figure A As the tectonic plates pull apart, cracks open up which become rift valleys.

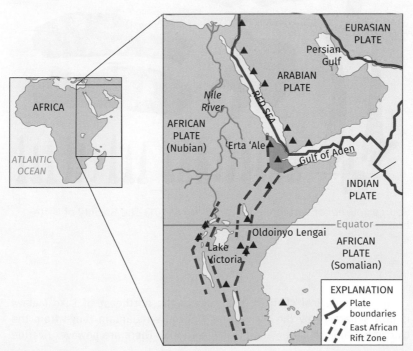

Figure B (a) Lake Malawi is between Tanzania, Mozambique, and Malawi in East Africa. (b) A selection of tiny fish fossils from the Chiwondo bed including teeth (B) from a tigerfish and fin spines (G) from cichlid fish.

(a)

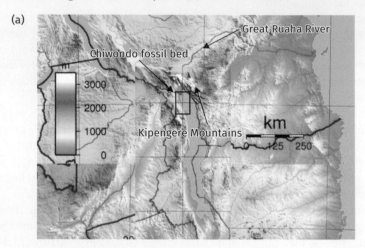

Source: Sadalmelik/Wikimedia

(b)

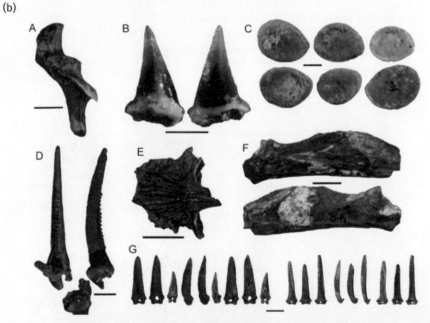

Source: *Journal of Vertebrate Paleontology* © 2013 The Society of Vertebrate Paleontology

Due to geological changes, this river to the northeast of Lake Malawi is now separated by the Livingstone/Kipengere mountain range from the catchment area of Lake Malawi (see Figure B a). There are however riverine

CS2.1 Table A This table indicates some of the major events in the African Great Lakes and cichlid speciation.

Event	Time line millions of years ago (Ma)							
	7	6	5	4	3	2	1	0
Lake Malawi formed		●——————●						
Deep water condition in Malawi				●—————————————·· —→				
— ·· — ·· — indicates fluctuating water levels.								
Divergence between Malawi cichlids and current cichlid species in Ruaha River		●			●			
Riverine deposits in Chiwondo fossil beds				●——●				
Divergent speciation in lake Malawi						●————————→		

deposits in the Chiwondo fossil beds along the northwestern side of Lake Malawi. These beds include fossils of non-cichlid fish species whose nearest living relatives live in the Ruaha river. This is evidence that rivers currently in the Malawi catchment were part of the Great Ruaha system before the Rift separated them.

Mitochondrial DNA evidence shows that the divergence of the various Malawi species from each other took place between 1.2 and 4.06 Ma. This fits in well with the fossil evidence of fish species in rivers around Malawi during that time and the fact that the newly formed Lake Malawi was first filled with water by 4.5Ma. It appears that ancestors of the Lake Malawi cichlids became isolated in rivers within the new Malawi catchment during the Pliocene when the Kipengere/Livingstone mountain range was formed. They then subsequently radiated into the lake itself—see CS2.1 Table A.

After the lake had formed, it experienced prolonged dry periods alternating with wet periods. During each dry period evolution will have continued in the small separated lakes—classic allopatric speciation. But when the dry eras ended and the lakes refilled, very similar races or subspecies will have come back into contact with each other. At this point it is likely that sexual selection (see Chapter 1) will have become more significant in driving speciation, and any further speciation will have been sympatric.

❓ Pause for thought

The huge diversity of cichlids in Lake Malawi is a product of the unusual volatility of the local environment due to the geology of the region. As the human population has grown, more land has come under cultivation and hill slopes have been deforested. Water for irrigation is taken from local rivers. How might these actions affect the lake and therefore cichlid diversity?

What is a species?

Perhaps the most revolutionary aspect of what Charles Darwin discovered is that species are not only fluid in time but also that they have rather fuzzy edges! Gone were the days of security when people knew that each and every species had its own well-defined place in Creation.

It was relatively easy to classify organisms as long as everyone thought of them as just 'being there'. It is human nature to name and group things. Aristotle (384–322 BC) classified animals and plants into groups and sub-groups based on similarities in shape. This is the 'morphological approach'. The discovery in about 1620 that lenses could be combined to make micro-scopes, enabling more minute observation, opened the floodgates of de-scriptions and classifications of plants.

Our modern classification system is based on that of Linnaeus (1707–1778) who created a hierarchy in his *Systema Naturae* enabling any living thing to be placed within a framework of progressively larger groups, from its genus and species up to its Kingdom (Animalia or Vegetabilia). His system was based on similarities in structure, from the very small differences which separated closely related species to the much larger differences which separated families, orders, classes, and phyla. In almost no time people began speculating on whether these similarities, like those in human families, could be traced in time. Within 20 years of Linnaeus' death, Charles Darwin's grandfather Erasmus Darwin was exploring the idea that species could change in his book *Zoonomia*.

This opened a Pandora's Box for would-be taxonomists. Once you accept that there is a lot of variability in a population, and that the likelihood is that it is slowly changing anyway, it becomes more difficult to base definitions on only the appearance, or morphology, of the individuals in a species. Over the last century or so a plethora of ways of identifying species have been suggested, including looking at similarities and differences in the proteome, mRNA, and the genome.

The morphological species model

Grouping organisms of similar appearance is the oldest method of classi-fying them—it gives us the morphological species. On a day-to-day basis, for most people, this is sufficient; but what happens when individuals come in different forms, as you saw in Chapter 1 with lazuli buntings, banded snails, and peppered moths? What about caterpillars and butter-flies, where the young look nothing like the adults? Grappling with these, and other complications, forced taxonomists to think more creatively.

Meanwhile, morphology is all that palaeontologists—scientists who study fossils—have to go on. This means that they are unable to take into account any differences the organisms displayed in life, such as colours of body coverings or behaviour.

Most of the classification system we now have is still based on morphology, but increasingly we use other tools as well to hone our species model.

Protecting genetic diversity in African elephants

Asian and African elephants have long been recognized as different as they have different body sizes and ear shapes. As African elephant populations have come under more and more pressure, it has become more important to understand their ecology in order to better manage and conserve them. Increased familiarity has led to scientists noticing consistent differences between African elephants that live in the savannah and forest-dwellers, with clear differences in their tusk shapes, and in both their ear shapes and sizes (see Figure 2.4).

DNA analysis from elephants in populations across Africa showed that not only are the differences between savannah and forest elephants more than half as big as the differences between the African and Asian species, but also that they are genetically more different than lions and tigers! Genetic analysis also showed very little interbreeding between forest and savannah elephants.

Now we have both morphological and genetic evidence to indicate that these elephants are separate species, although the fact that they can produce fertile hybrids means some scientists still consider them both to be subspecies of *Loxodonta africana*. The forest elephants mostly live in very dense forests, often in politically unstable countries with high levels of ivory poaching, so it is difficult to count and conserve them. We may lose one of these precious species before we have even finalized our biological understanding of how it has evolved.

Figure 2.4 Forest elephants (left) are smaller, have more rounded ears, a different skull shape, and straighter more downward-pointing tusks than the savannah elephant (right).

Source: Benh Lieu Song (forest elephant) and Thomas Breuer (savannah elephant)

The biological species model

Even to scientists applying morphological criteria, it was obvious that the males and females of many species differ. But they could also see that these apparently very different animals were able to breed successfully: this led to Ernst Mayr's 1942 definition that species are groups of actually or potentially interbreeding natural populations, which are reproductively isolated from other such groups. He included 'potentially' since many species are spread over a wide geographical area. This persists as what is known as the biological species definition.

The opposite situation exists too. Morphologically almost identical species may be shown to be reproductively incompatible when individuals from separate populations are mated. These species are called cryptic species, since the fact that they are different species is hidden from view.

Copepods: foundation of the sea's food web

Copepods are tiny crustaceans only 1–2mm long which form the bulk of plankton, and so are the main food source at the bottom of both marine and freshwater food webs all over the world. Understanding their ecology is vital to food security, as the juveniles of many of the marine fish species we eat depend upon copepods for food. For example, all the copepods in Chesapeake Bay, off the coast of Maryland and Virginia, look the same, so are they all the same species?

Scientists placed copepods from saltwater and freshwater areas of the Bay together in breeding tanks in various combinations. The number of eggs produced from between-types crosses was not significantly different from same-type crosses. However, only 2% of the eggs produced by between-type crosses hatched, compared to about 50% of the eggs produced by same-type crosses. Combined with other data such as size, this persuaded the team that the fresh and saltwater copepod populations are different species.

Reproductive isolation: how species stay distinct

Once speciation has occurred, species must retain their genetic separateness even when inhabiting the same habitats if they are to remain a separate species. This is usually achieved as a result of reproductive isolation. This reproductive isolation can happen at various stages of the reproductive process, and is either pre-zygotic or post-zygotic.

Pre-zygotic reproductive isolation refers to any barrier that prevents the egg being fertilized by sperm from another species. This could be:

1) temporal or habitat isolation—eg the toads *Bufo americanus* and *Bufo fowleri* have different mating seasons

2) mechanical isolation eg incompatibility of male and female anatomies or adaptation of flower shape to specific pollinators in plants (see Figure 2.5).

3) behavioural isolation eg the flashing sequence of male fireflies are species-specific and so only attract the right female

4) gametic isolation eg a pollen grain will only grow a pollen tube through the style of the right species of plant, gamete release is synchronized in whole reefs, and yet the corals do not hybridize

Post-zygotic reproductive isolation refers to mechanisms which prevent a hybrid surviving or being fertile. The copepod example above, where fertilization of the female led to the production of non-viable eggs, is an example of post-zygotic reproductive isolation. The formation of infertile hybrids as a result of interbreeding between species is another form of post-zygotic reproductive isolation. Infertile mules, which result from a cross between a donkey and a horse, are a classic example of this process.

Figure 2.5 The shape and scent of early spider orchids lure male solitary mining bees into trying to mate with them, pollinating them at the same time. © Anthony Short

Human interventions

Speciation is a key aspect of evolution. Perhaps unsurprisingly, we humans got involved long before we had any understanding of either process. A good example is the story of wheat. Almost all the wheat eaten in the world is bread wheat (*Triticum aestivum*), but other older species continue to be cultivated by farmers growing for local markets in small villages. These include ancient wheats such as einkorn and spelt. Modern bread wheat has been selectively bred, and the journey began over 10,000 years ago. Scientists are now mining the genomes of some of the more ancient wheat species for useful genes to help our modern hybrids cope as the global climate changes. One example is the salt tolerance gene from einkorn (*Triticum monococcum*) which Australian scientists have introduced into the durum wheat (*Triticum durum*) used for pasta.

Extinctions: the end can be the beginning

The scene is romantic and exotic. An intrepid geological explorer is exploring mountains deep in the uncharted wilderness and is leading his horse along a small precipitous mountain track. His horse trips, and as he catches himself from falling down the fathomless scree below he spots the most extraordinary fossil he's ever seen...

The Burgess Shale is its own legend. It owes its existence to a series of random slips of mud off a precipitous underwater cliff in a sea in the unimaginably distant past. The slips were disasters to those inhabiting it and whole communities were wiped out. The explorer who discovered them millions of years later was palaeontologist Charles Doolittle Walcott—and he clearly knew almost immediately that he had found something very, very special, even if he did rather exaggerate the story of its discovery!

The Burgess Shale fossils continue to provide revolutionary insights into how evolution progresses, even more than a century after their discovery.

It is one of an elite collection of fossil beds known as lagerstätten due to their extraordinary richness of well-preserved fossils, and it is legendary because it is a snapshot of the progress of the Cambrian Explosion (see Figure 2.6). Walcott never completed his formal education but he had a tremendous passion for fossils. This eventually led him to lead the prestigious Smithsonian Institution. He discovered the Burgess Shale fossils while on a summer fieldwork break when he was almost 60, and he returned every year, with his family acting as assistants, until he was 67. They collected over 80,000 fossils and amassed a huge collection of panoramic photographs of the Canadian Rockies, where the fossil beds lie.

What Walcott did not know was that his finds would help re-write our understanding of evolution. They confound our expectations because they contain organisms so different from one another that palaeontologists are compelled to place them into at least fifteen different phyla, only seven of which exist today.

When Harry Whittington set out to review the Burgess fossils in the 1970s he expected to take no more than a year to describe all the arthropods—but ended up taking four and a half years just to write his monograph on *Marella*, which Walcott had described as a primitive trilobite. In the end Whittington had to create a new arthropod class because *Marella*'s characteristics did not comfortably fit in any known group.

By the time he got to the much rarer *Opabinia* (see Figure 2.7), the painstaking dissection methods he had developed came into their own as he dissected through the animal's carapace to seek the arthropod limbs which 'should' be there. They were not, and neither did it have the mouthparts

Figure 2.6 Just one of the amazing organisms from the Burgess Shale. The incredible detail results from the very fine sediments in which they were preserved.

Source: Fu, D., Tong, G., Dai, T., Liu, W., Yang, Y., Zhang, Y., Cui, L., Li, L., Yun, H., Wu, Y., Sun, A., Liu, C., Pei, W., Gaines, R.R., and Zhang, X. (2019). The Qingjiang biota—A Burgess Shale-type fossil Lagerstätte from the early Cambrian of South China. *Science* 22 Mar 2019: 1338–1342.

Figure 2.7 This reconstruction of *Opabinia* shows how it is segmented like an annelid, but has eyes and a carapace like an arthropod.

characteristic of arthropods. In the end Whittington was unable to place it into any modern phylum—a new one had to be created.

According to the conventional view, variety and complexity increase over time. The picture painted by the Burgess Shale seems to turn this view upside down. It appears that in a world devoid of animal life there were limitless niches which created opportunities for evolutionary experimental life forms. During that first outburst of diversification, a mind-blowing number of bizarre anatomical possibilities arose. Today, despite a further 500 million years of evolution, we have only about 32 animal phyla. The massively overgrown tree of life which resulted from the Cambrian Explosion was tested to destruction and severely pruned over subsequent eras.

You'll discover more about the Burgess Shale fossils and what they have shown us in **Scientific Approach 3.1**.

Catastrophic floods and extinction

The first recorded Cambrian fossils, discovered in 1698, were the arthropod trilobites. The science of palaeontology was established by Georges Cuvier in the late 18th century, when he published papers comparing the anatomy of fossils and living species which were similar. He also established that extinctions—the loss of an entire species—can happen, with two papers in which he compared the anatomy of living (extant) and fossil elephants and sloths. Extinction had been considered by many to be merely controversial speculation, as there was no evidence to support the theory. In his *Essay on the Theory of the Earth* (1813) Cuvier proposed that the reason fossils of now-extinct species were found in sedimentary rocks was that they had been wiped out by periodic catastrophic flooding events.

By the 1830s leading geologists were using fossils to date rock strata. However, there were a variety of theories regarding the origin of species with a minority favouring the theory of transmutation of species (what we now call evolution). Others, bemused perhaps by the sudden jumps which can be observed in the fossil record between organism types, interpreted the same fossil record as sequential Creation events over geological time.

This included Darwin until after his return from the Galapagos. Darwin acknowledged that the lack of fossils pre-dating the Cambrian trilobites, and other gaps in museum collections, was a drawback to the theory of evolution but explained it as follows:

> *"I can answer these questions and grave objections only on the supposition that the geological record is far more imperfect than most geologists believe. ... The number of specimens in all our museums is absolutely as nothing compared with the countless generations of countless species which certainly have existed."*

Extinctions have always been an essential part of evolution (see Figure 2.8). Groups of species which have monopolized various niches often disappear, leaving those niches wide open to be exploited again. There have been numerous local extinctions, such as might happen if a lake dries up or a coastal cliff is elevated by seismic movements. More perplexing have been the five major extinction events, when the majority of species present globally at the time have disappeared.

Figure 2.8 It took over 2000 million years for eukaryotes to arrive after the first cells had evolved. And yet it is only 541 million years since the first outburst of diversity at the Cambrian Explosion.

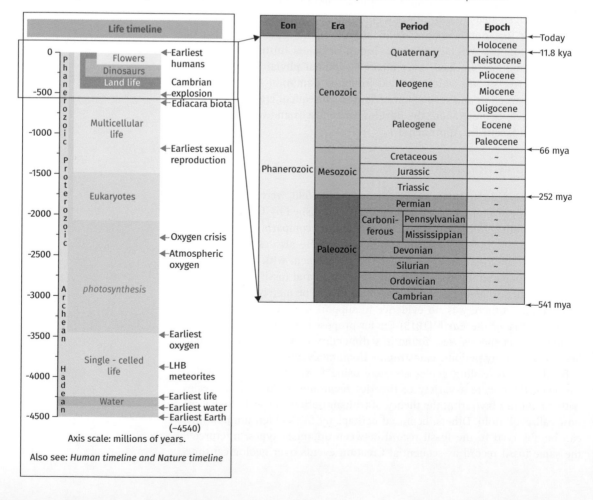

Being human, we tend to think in terms of the human lifespan, but the 'blink of an eye' in geological time is often millions of years. This is because the sediments that coalesce into stone accumulate very slowly, maybe only a few millimetres in a year. Marine fossils are more numerous than terrestrial ones because the organisms were living an aquatic existence, where sediment build-up is a continuous occurrence. For this reason extinction events are generally defined primarily by marine species.

The crucible of creation

The major extinction events are periods of time when a large number of species disappeared across the globe; we now think the most likely cause of these events were catastrophic changes in the climate. However, we know from studying the biochemistry of minerals in the Earth's crust that there was another massive event at the dawn of Life itself. This event is not generally described as an extinction event, because it has been impossible to ascertain exactly what was alive then. But it is of such earth-shattering importance that it has been described as a Catastrophe, Crisis, Holocaust, and Revolution: it is the Great Oxidation Event.

The oxygen revolution and the dawn of eukaryotes

The seeds of this oxygen revolution were sown 3.5 billion years ago, probably due to the innovation of photosynthesis in small primitive microbes called cyanobacteria. Their photosynthetic descendants still exist, often forming films on aquatic surfaces and building larger structures called

Figure 2.9 By studying the ecology and physiology of modern day stromatolites, we can infer a lot about ancient environments when we find fossils of stromatolites in surprising places including deserts and mountain tops.

Source: Juanmoro/Istock

stromatolites (see Figure 2.9) in shallow water. Today these microbial mats contain complex communities of interdependent microbes including archaea, bacteria, algae, fungi, and protozoans. Research into their distinctive chemistry has given us insights which geologists use to interpret the anomalous structures, sometimes known as Microbially Induced Sedimentary Structures (MISS), found in sedimentary rocks. These distinctive microscopic structures contain bound sand grains and characteristic minerals such as pyrite (iron sulphide) which is manufactured by chemosynthetic bacteria in the absence of oxygen.

The evolutionary breakthrough came when bacteria evolved a way of using light to fix carbon dioxide. Oxygen was the waste product—and it was toxic for many organisms. As the available oxygen in the environment increased, it was only a matter of time before a bacterial species evolved the ability to respire aerobically.

The theory is that, over time, some of these aerobic bacteria survived endocytosis by a larger unicellular organism, and retained the ability to respire aerobically whilst living inside the larger cell. Having endosymbiotic bacteria respiring aerobically, producing a relatively large source of ATP from each glucose molecule, would give an organism a considerable evolutionary advantage at a time when other species were relying on glycolysis and fermentation—both anaerobic processes which release far less energy when carbon compounds are respired.

The endosymbiotic hypothesis suggests these aerobic bacteria became prototype mitochondria in the new eukaryotic cells, reproducing independently as the cells replicated, so each new cell had its own bacterial 'mitochondria'. Modern mitochondria continue to have their own DNA and to replicate independently of the nuclear DNA. The increased efficiency of aerobic respiration enabled cells to evolve storage molecules such as lipids. The cell membranes of modern eukaryotes contain lipids called sterols, the most familiar one being cholesterol.

By 2.1 billion years ago, the multicellular photosynthetic eukaryotic alga *Grypania spiralis* was thriving all over the world. This was one of many times that multicellularity evolved. It took more than another billion years for multicellular animals to evolve, and another 500 million years for complex organisms to appear during the Ediacaran period. These organisms are truly alien to us: all soft-bodied with body plans with defy classification into modern taxonomic groups (see Figure 2.10). The stage was set for the Cambrian Explosion, a mind-boggling pageant of adaptive radiation and diversification which sowed the seeds of today's phyla.

The Burgess Shale is one of the many geological strata that have accumulated to create today's Rocky Mountains. Each was once a surface, and so its nature is affected by where it was on the planet, and whether it was a terrestrial or aquatic surface. Scientists analyse the deposits to discover more about the climate, environment, and location in which each layer was created. There are many events which could change the composition of the deposits being laid down.

Each change is marked by a change in colour or texture (see Figure 2.11). When a palaeontologist finds a transition such as one of these, it is important to decide whether it was due to a local event, such as an earthquake, or a global event such as an ice age. Over the last 200 years scientists have gradually built up a picture of the history of life during which there have been five occasions when the biosphere has been decimated by global events.

Figure 2.10 What is it? Specimens of the ediacaran *Dickinsonia* range from a few millimetres to 1.4m long but are rarely more than a few millimetres thick.

Source: Universal Images Group North America LLC/DeAgostini/Alamy Stock Photo

Figure 2.11 Geological strata on the Jurassic coast of Dorset, with pale bands of dolomite separating layers of fossil-rich shale. Each band represents a different kind of deposit, laid down in different conditions. © Anthony Short

Five major extinction events

The traditional Darwinian evolution story is one of steady change leading to adaptation, where individuals may die because they are relatively less well equipped than others to pass on their genes to the next generation. Extinction events occur when something extraordinary happens, something for which significant numbers not only of individuals but also entire species were unprepared. The best known of these extinction events in the public mind is when the dinosaurs became extinct. Individual species disappear frequently, and there are extinctions of many sizes—extinction is a key feature of evolution. But there have been five major extinction events when at least two thirds of the species known to be alive at the time disappeared from the global fossil record. Scientists are convinced that each must have involved a sequence of events which led to a planetary change.

- **End of the Ordovician, 444 million years ago, 86% of species lost:**

Life had begun in the warm, sunlit, shallow seas with most of the land in the supercontinent Gondwana. By the end of the Ordovician Radiation, most modern marine invertebrate phyla, together with fish, existed in the oceans, and plants and arthropods began to invade land. Throughout the Ordovician, Gondwana shifted towards the South Pole and much of it was submerged underwater. The final million years of the Ordovician was marked by a major glaciation, which cooled and shrank the marine habitats as sea levels dropped, causing ecological disruption and mass extinctions.

- **Late Devonian, 375 million years ago, 75% of species lost over about 20 million years**

There were a number of extinction events during this time, probably due to a number of abiotic factors. Asteroid impacts can cause volcanic eruptions which release huge amounts of CO_2 and accentuate global warming. Aquatic organisms were wiped out as oxygen levels in the water plummeted due either to eutrophication following algal blooms, or a reduced mixing of water layers as oceans warmed—or a combination of the two. Trilobites, the most diverse and abundant of the animals that appeared in the Cambrian explosion, survived the first great extinction but were nearly wiped out this time (see Figure 2.12).

Figure 2.12 Trilobites teetered on the brink, after millions of years of success. © Anthony Short

- **End of the Permian, 251 million years ago, 96% of marine species lost and over 70% land species**

Known as 'the Great Dying', this was by far the worst extinction event so far. In the space of just 60,000 years it nearly ended life on Earth. The abrupt transition in rocks at the Permian–Triassic boundary, from being filled with fossilized pollen to almost no pollen but lots of fossilized fungi suggests food chains collapsed as most plants died and decayed.

It began when volcanic eruptions across Siberia deposited more than 720,000 cubic miles of lava across the world, and released at least 14.4 trillion tons of carbon—more than twice the amount which would be unleashed if all the fossil fuel on Earth were burned. Methanogenic archaebacteria converted this carbon dioxide to methane, and global temperatures surged by about 10 ºC. Sulphur and nitrogen oxides in the volcanic gases made acid rain. As the oceans warmed, the currents which kept the waters mixed stalled, and levels of dissolved oxygen dropped. The warmth increased the metabolic rates of marine organisms, and so they needed more oxygen just when less was available. The results were disastrous.

- **End of the Triassic, 200 million years ago, 80% of species lost**

This extinction event, over a period of only about 10,000 years, is the most enigmatic of the major extinctions. Like the other major events, it was global, but the jury is out over whether the primary cause was volcanic, gradual climate change, or a meteorite impact. It could be due to a perfect storm of individual events which each contributed, when extensive volcanic eruptions accompanied the splitting of Pangaea to form the Atlantic Ocean. Once again the Earth warmed considerably as atmospheric CO_2 levels quadrupled. This may have acidified the oceans, making it more difficult for marine creatures to build their calcium carbonate shells.

Typical features of this period are large bone beds such as the mudstone in Loulé, Portugal, where hundreds of amphibians perished as a lake dried up. Terrestrial ecological niches were vacated as the crocodilians, the dominant vertebrate group, died out. This allowed dinosaurs to assume dominance during the Jurassic period.

- **End Cretaceous, 66 million years ago, 76% of all species lost**

The Cretaceous was marked by a warm climate and a high sea level. Large shallow seas provided a rich habitat. The iconic ammonites (see Figures B and D in SA 2.1) had diversified to fill a global multitude of marine niches. These molluscs, early relations of squids, evolved rapidly so that each species had a relatively short lifespan, making them very useful for dating sedimentary rocks. Meanwhile, dinosaurs of all shapes and sizes ruled the land.

Then one day, about 66 million years ago, an asteroid slammed into the sea at 20km/second off what is now Mexico's Yucatán Peninsula, leaving a crater 150km across and 20km deep. A mega-tsunami, around 1.5km high, drowned surrounding coasts. Vast quantities of dust and sulphate aerosols were thrown into the atmosphere, bringing on severe global cooling. Heavier debris cast into the upper atmosphere became incandescent upon re-entry, igniting wildfires, while the sulphates dissolved in rain and acidified the oceans. Overnight, the ecosystems that supported non-avian dinosaurs began to collapse, and the final great extinction event so far took hold.

Scientific approach 2.1
How understanding ecology helps us to interpret the fossil record

The science of ecology was in its infancy when Walcott discovered the Burgess Shale fossils. Today we have a clearer understanding of how ecological communities are made up, and the exact role of each organism within them. Modern taxonomists can therefore look for structural or chemical signatures in fossils which might indicate the organism's niche while it was alive. This in turn can help us to understand how the environment may have changed at an extinction event, and therefore hypothesize the causes of it.

For many, the dinosaur extinction at the end of the Cretaceous is the most iconic extinction. However, when the meteorite impact hypothesis was first proposed it was not taken seriously. It seemed too far-fetched. Nevertheless, scientists began finding pieces of fossil evidence which supported the idea. An example is the discovery that many species which depended on photosynthesis, including phytoplankton and land plants, declined or became extinct. This would make sense if atmospheric particles from an impact were blocking light from the Sun reaching the Earth.

Evidence from many sources—from the rock strata in the giant crater discovered in Mexico to all the fossil stories which emerge from the timeframe—supports the meteorite hypothesis for the Cretacean extinction event.

Fossils tell us not only about the organism fossilized, but also its predators. Herbivores and carnivores may leave evidence as damage on fossils. Figure A shows how the types of herbivorous insects present can be determined by microscopic examination of damaged fossil plant material. A striking discovery about the end of the Cretaceous is that it seems no purely carnivorous or herbivorous species survived. Many of the survivors were carrion-eaters, omnivores, and insectivores—best able to survive in a world with few plants and many carcasses!

Fossil teeth and other features are invaluable in determining diet (see Figures B and C). Sometimes teeth can be found which actually fit holes in

Figure A Some leaf fossils are so well preserved that scientists can identify both the type of plant and the type of insect that fed upon it.

10 mm

Figure B Did a predator leave this distinctive pattern of damage when it bit the ammonite to extract the animal from inside? © Jane Still (from the Etches Collection, Kimmeridge, Dorset)

Figure C Are these aptychi, found with ammonites in the Kimmeridge Clays, a primitive beak for feeding like those seen in cephalopod molluscs today, or an operculum for protection? © Anthony Short

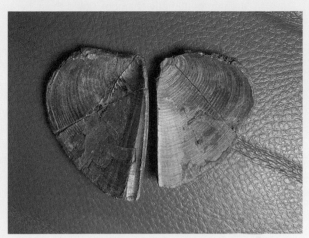

other fossils! The ammonite in Figure B is typical of the Kimmeridge Clays. The damage seen is always in the same place. Was it caused by a particular type of predator which bit the ammonite to extract the animal from inside?

Other aspects of a fossil's body shape, together with studying local fossilized spores and pollen, can help place it in the type of habitat it may have lived in. For aquatic organisms, the nature of surrounding sediments indicates which type of watery environment it lived in.

Even life cycles can be inferred from fossils. Palaeontologists had been mystified by what they thought were two closely related ammonite species, one with a horn on its shell and another larger species with no horn (Figure D). However, the consistency of the sizes, together with the fact that the two types of fossils were always found together, led to the conclusion they were

Figure D One region of the Kimmeridge Clays contained only ammonites and cephalopods. Any eggs must belong to one of them! Jane Still (from the Etches Collection, Kimmeridge, Dorset)

Male   Female   Egg cluster

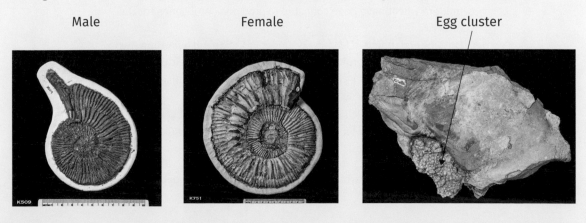

a sexually dimorphic species. Comparing them with modern relatives led to the conclusion that the smaller horned specimens are males. Fossilized egg clusters (see Figure D), similar to those laid by modern cuttlefish, a distant relation of ammonites, were found on the ancient mud surface as well as in an ammonite body chamber. Could these be ammonite eggs? Maybe one day a micro-CT scan will show us what is inside!

 Pause for thought

How might applying knowledge of the ecology and interrelationships of modern species help us to better understand the ancient environments in which particular fossils were laid down? Suggest ways in which a better understanding of the climate changes of the past might help us to create better models for understanding future climate change and its effect on the biosphere.

 Chapter Summary

- Reproductive (genetic) isolation is necessary for speciation to take place.
- Allopatric speciation occurs when a species becomes geographically separated. Sympatric speciation occurs when a species diverges into more than one species in one location.
- The huge variety of African cichlid fish may be largely due to repeated allopatric speciation events over geological time. The variety of colours may be an example of sympatric speciation due to sexual selection accentuating small variations in morphology and colour.
- Speciation may occur as a result of hybridization. Modern bread and durum wheat are examples of this. Older varieties and ancestors of modern wheat have been collected to provide useful genes for the future.
- Our understanding of what a species is has changed over time. Two concepts are the morphological species and the biological species.
- Extinctions are a necessary part of evolution as they cause niches to be vacated so a new set of speciations can refill them.
- Photosynthesis was an evolutionary breakthrough as it gave organisms access to light as a source of energy, and eventually access to oxygen.
- Oxygen enabled the evolution of eukaryotic organisms, which led to the evolution of heterotrophic organisms such as animals.

- Extinctions happen constantly but there appear to have been five major worldwide extinctions, which have occurred as a result of planetary changes.
- Studying the biochemistry, physiology, anatomy, and ecology of modern organisms is helping us to better understand both the organisms whose fossils we find and the environments they were living in.

Further Reading

https://www.awf.org/news/africas-forest-elephants-called-separate-species#:~:text=Forest%20elephants%20are%20smaller%20than,also%20differs%20between%20the%20two

Using DNA to identify speciation in elephants

http://html5.ens-lyon.fr/Acces/FormaVie/20130304/Avraham_Levy/audio.html#diapo004

Narrated Powerpoint about the evolution of wheat

https://burgess-shale.rom.on.ca/en/

The Burgess Shales

http://www.palaeocast.com/episode-48-the-burgess-shale/?fbclid=IwAR3wOmFPN1JOYhOIqkDiBKLguBM0DUVae4qt86Pl0hNXKTE0flNJdJpEWJY

Podcasts about Palaeontology, eg Burgess Shale

www.nationalgeographic.org/topics/resource-library-evolution/?q-&page=1&per_page=25 https://www.nationalgeographic.com/science/prehistoric-world/permian-extinction/

The National Geographic has good evolution articles and videos, eg

Discussion Questions

2.1 Legends are usually based on an historical fact, but certain aspects of the story may have become exaggerated in order to make it a more inspiring story. To what extent is the story of the discovery of evolution a legend? Suggest other stories of scientific discovery which could be described as legends.

2.2 Some scientists question whether the Late Devonian mass extinction 'counts', as it took place over a period in excess of 20 million years. What makes an 'extinction event'?

2.3 Many believe that we are in the middle of the sixth mass extinction. Discuss this idea, gathering evidence to support arguments on both sides.

3 WHAT'S THE EVIDENCE?

We saw in Chapter 1 that every species that has existed on Earth can tell a story about its journey from the Last Universal Common Ancestor (LUCA), about 3.8 billion years ago, to the present day. We can think of these stories as being like the individual fibres in a rope, all interconnected and twisted around each other. The stories are told using countless small pieces of circumstantial evidence.

We shall see in Chapter 4 that the importance of each piece of evidence is fiercely discussed and debated; often arguments develop that become quite intense. Yet, just as each fibre contributes to the strength of the whole rope, so each small piece of evidence contributes to the strength of the theory of evolution. The principles discovered by Charles Darwin, presented in the earlier chapters, have been shown to be correct time and again, and scientists have great confidence that evolution is a reliable model of the development of life on Earth.

This chapter considers the types of evidence that are used to show how evolution has happened. In doing so, we will glimpse parts of the evolutionary journeys that a few species have made. We could spend many lifetimes trying to understand the whole of these journeys (see Figure 3.1), and so this chapter gives only the briefest of introductions.

Figure 3.1 Tuatara, found in New Zealand, are reptile-like animals. Their evolutionary story began before the dinosaurs, and continues today, although their numbers are now tiny and threatened. © Anthony Short

The evidence from fossils

Fossils are the most familiar evidence for evolution—and they provide rich, varied, preserved evidence of life in past geological ages. A fossil may be preserved bones or feathers, the exoskeletons of arthropods, or the shells produced by molluscs (Figure 3.2). They may be footprints, leaving a poignant trace of journeys back in time, or eggs and embryos of animals which never made it into life, or fossilized faeces. There are plant fossils, from tree trunks (Figure 3.2) and leaves to pollen grains and seeds. Fossils may be preserved in rock, or amber, or permafrost. They may be whole organisms, or simply imprints of where an organism once grew. There is even, in rare cases, fossil DNA.

The Jurassic Coast, that forms part of Dorset and East Devon in England, is a World Heritage site because its rocks contain many fossils, formed between 250 and 66 mya in the Triassic, Jurassic, and Cretaceous periods. These rocks enabled Mary Anning, one of the earliest and best-known fossil hunters, to make a living collecting and selling fossils to Victorian scientists across Europe (see Figure 3.3). Case study 3.1 tells you more about this amazing woman.

Figure 3.2 From fossil shells (a) to fossil trees (b), all types of fossils offer us a glimpse back in evolutionary time. © Anthony Short

Figure 3.3 A plesiosaur, collected by Mary Anning and still displayed in the Natural History Museum, London, UK.

Source: Salajean/Shutterstock

Case study 3.1
Mary Anning—fossil hunter extraordinaire

Mary Anning was really Mary mark II—her four-year-old sister Mary burned to death six months before her birth, and her distraught parents named their new daughter after the child they had lost. Then, when Mary mark II was one, the nurse carrying her was struck dead by lightning, and the baby was found unconscious. When Mary recovered, she was bright, lively, and interested in everything—before the lightning strike she was apparently a rather dull baby!

The Anning family lived at Lyme Regis in Dorset, UK. They were poor, and made extra money by collecting curios (what we now call fossils) from the beach and selling them. Mary's father died when she was twelve and only two of the eight Anning children survived their life of poverty. But Mary and her brother Joseph doggedly continued collecting fossils and selling them—one woman paid half a crown (12.5p) for a beautiful ammonite, a week's food for the family, so it was worth doing!

Then in 1811, Joseph unearthed the giant head of a fossilized creature with an enormous eye—they thought it was an enormous crocodile. Mudslides covered the area around the skull, but a year or so later Mary found the entire backbone—60 vertebrae—of the amazing creature. It was about 5 metres long. The skeleton was bought by the local Lord of the Manor for £23—food for six months!

Figure A Mary Anning on the Purbeck coast where she found so many of her specimens—note the geological hammer in her hand.

In the years to come, Mary Anning uncovered fossils of all sorts, including many of her 'crocodiles'. Scientists called them ichthyosaurs—fish-lizards—because while they had many lizard-like features, particularly in the teeth and jaw, some of their bones were fish-like. Most of the 'gentlemen geologists' who emerged with the new science were happy to visit Dorset, and use this dedicated and talented young woman to supply them with specimens, or at the very least show them the right places to search. Yet when scientific papers about ichthyosaurs were published, Mary Anning was given no recognition. She wasn't even sent copies of the papers or allowed to name her finds.

However, there was a time when Mary's luck ran out—however hard she tried she couldn't find any more major fossils. The family was slipping back into poverty when one of the scientists they had helped—Lieutenant-Colonel Thomas Birch—decided to sell the fossil *Ichthyosaurus* he had got from them, and send the Annings the proceeds.

The money enabled the family—Mary, her mother, and her brother—to keep going until their luck turned again. Mary discovered several different species of ichthyosaurs, and then in 1823 she uncovered something she had never seen before in all her years of fossil hunting. It was another major discovery—an almost complete plesiosaur skeleton (see Figure 3.3). It caused great excitement and the Duke of Buckingham offered £200.00 for it. There was a bit of bother when Baron Georges Cuvier, a highly respected French expert, said that such an animal could never have existed because its neck was too long and thin—but his opinion was soon overruled as other findings confirmed Mary's new beast. Mary Anning didn't just find the fossils—she excavated them and then mounted them in plaster in wooden frames for display, eventually becoming known as 'the most eminent woman fossilist'. Although some found her proud and full of her own opinions, locally she was regarded as kind and supportive of anyone struggling with poverty, as she had done.

In 1828, just as the Annings were yet again short of money, Mary made another amazing discovery—the first flying reptile or pterodactyl to be found in Britain. But in spite of all her efforts, the Anning family often found it hard to make a living, especially after Mary invested her life savings with a con-man who ran off with the lot in the late 1830s. Eventually she was granted a pension of £25 per year which saved her from starvation!

In 1842, Mary's mother died, and Mary herself began to suffer terribly with breast cancer. She died in 1847 and was buried in Lyme churchyard. She was given a eulogy at the Geographical Society by the President of the Society himself—very rare indeed for someone who was not a Fellow of the Society—and a stained glass window in her honour was placed in the church at Lyme. The inscription reads:

> This window is sacred to the memory of Mary Anning of this parish in commemoration of her usefulness in furthering the science of geology, as also of her benevolence of heart and integrity of life.

As a footnote to this story, in 2020, fossil bones from a new species of dinosaur, related to *Tyrannosaurus rex*, were found on the Isle of Wight, UK by three different amateur fossil hunters. The bones have been studied by scientists at the University of Southampton—and the names of the people who found the bones are on the scientific paper as co-authors.

❓ Pause for thought

Suggest reasons why Mary Anning received so little recognition for her work during her lifetime. How much have things changed—and why?

How are fossils made?

The fossils most often found in rocks are the remains of the hardest parts of animals, which already contained minerals, such as shells, teeth, and bones and the hard parts of plants, such as wood containing lignin. The process of fossilization is summarized in Figure 3.4.

Occasionally, special circumstances mean that fossils of soft-bodied animals and flowers are preserved. The seas of the Burgess Shale in Canada (≈508 million years old) provided suitable conditions for the formation of rocks that preserved many of the soft-bodied marine animals associated with the Cambrian explosion discussed in Chapter 2.

Figure 3.4 Summary of the process of fossilization.

The dinosaur dies in a river.

The body is covered with sediment.
The meat decomposes.
The dinosaur becomes a fossil.

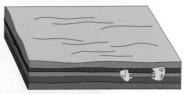

The sediments become rock.
The skeleton is pressed.

The earth's movements
raise the layers
of the rocks to the surface.

The rock erodes,
exposing the fossil.

Scientific approach 3.1
Developing techniques to study fossils of soft-bodied organisms

This is a guest contribution by Professor Derek Briggs, who worked on the fossils from the Burgess Shale for his PhD in 1976. He is currently the G. Evelyn Hutchinson Professor of Earth and Planetary Sciences at Yale University and Curator of Invertebrate Paleontology at Yale's Peabody Museum of Natural History.

We rely on fossils to document the history of life on Earth. Reconstructions and animations of fossils with modern computer graphics may give the impression that everything is known, but fossilization is a fickle process. Animals with hard parts (biomineralized shells, bones, or teeth), which have the best chance of being preserved, generally account for less than 40% of the different animals in a modern marine assemblage—the rest are soft-bodied and would normally decay giving them little chance of becoming fossilized. Thus there is a limited fossil record of a whole range of extinct soft-bodied creatures, particularly invertebrates, which were very different to the familiar animals of today.

Figure A Professor Derek Briggs on Mount Stephen in the Canadian Rockies. Walcott's famous Burgess Shale quarry is on the snow-covered ridge in the background to the right.

(a)

The major animal body plans, from various worms to vertebrates, became established during one of the most important events in the history of life, the so-called Cambrian Explosion. This steep increase in diversity took place during the Cambrian Period, 541–485 mya. A combination of conditions favoured the preservation of soft-bodied animals during this period: the prevalence of clay minerals that inhibited decay in the sediment, rapid cementation which sealed in carcasses soon after they were buried, and limited disturbance by burrowing animals. The result was a number of famous fossil deposits such as the Chengjiang biota of Yunnan Province in China and the Burgess Shale of British Columbia in Canada, which preserve a range of remarkable extinct forms that hold important clues to the early evolutionary history of today's animals. These include the lobe-limbed *Hallucigenia* with its paired dorsal spines, *Opabinia* with its five eyes and long proboscis, and the giant predator *Anomalocaris*, all of which are offshoots of the lineage leading to the most diverse of living groups, the arthropods.

Research on remarkably preserved fossils presents a number of challenges. Localities are rare and may be difficult to access—the Burgess Shale, for example, is high in the Canadian Rockies in a national park and special permission is required, even to visit. Collecting involves heavy physical work and care must be taken when splitting the rock to retain both halves of the specimens. A record needs to be kept of precisely which layer fossils come from so that associations can be used in interpreting their ecology. Once back in the laboratory, fossils are prepared for study; it may be necessary to remove thin layers of rock with a microengraver or modified dental drill or, in some cases, an airbrasive apparatus which acts like a miniature sand blaster.

Even though evidence of soft tissues survives, the carcass may have under-gone some decay prior to fossilization resulting in the loss or distortion of some features. Fossils may be flattened in the rock in different orientations depending on how they were transported and buried by sediment—different specimens provide different information about the animal.

Working with soft-bodied fossils requires a detailed understanding of how they became fossilized. Some arthropods from the Chengjiang biota, for example, including fine details of the limbs, are preserved in pyrite. This iron mineral is denser than the sediment surrounding it and can be imaged using high resolution X-ray computed tomography, commonly referred to as CT scanning. This method reveals the fossil in three dimensions so that it can be manipulated and dissected on a computer and printed, magnified if necessary, with a 3-D printer. Burgess Shale specimens are preserved as organic remains, which have transformed to more stable carbonaceous compounds over time. Features such as the digestive and nervous systems may be evident because they were the site of mineral formation during fossilization.

Fossils are examined with a microscope. The interpretation of complex morphology may be aided by making detailed outline drawings using an apparatus called a camera lucida which employs mirrors to project an image of the specimen under the microscope onto a sheet of paper. If fine structural details have survived they can be observed with a scanning electron microscope. It might be assumed that the carbonaceous component of fossils as old as Cambrian is so altered that evidence of their original composition is largely obliterated. However, Raman spectroscopy, which uses laser light to identify the nature of chemical bonds in a sample, shows promise in detecting chemical 'fingerprints' that may help to identify unusual soft-bodied fossils. Different minerals may be characteristic of particular anatomical features, allowing them to be reconstructed and interpreted. The nature of these minerals is generally inferred from maps showing the distribution of chemical elements on a specimen which can be obtained by energy-dispersive X-ray spectroscopy without removing samples for analysis. A new non-destructive method (in situ selected area X-ray diffraction) allows the identity of the minerals to be determined directly and more accurately rather than relying on element maps. Thus, chemical analyses can contribute in important ways to our understanding of soft-bodied fossils.

Early soft-bodied fossils often represent strange extinct forms separated by hundreds of millions of years from their nearest living relatives. Determining their role in the evolutionary history of the group relies on reconstructing a branching diagram of relationships known as a phylogenetic tree. This is a complex process which involves coding multiple anatomical features (i.e. characters such as number of segments, presence/absence of eyes) for many different animals. These data are then analysed using computer programs which apply different methods based on criteria such as parsimony (seeking the shortest tree) or probability models. Gene sequences may be incorporated where the relationships of fossils and their living relatives are the target although they are clearly not available for ancient fossils.

Where soft-bodied fossils are preserved they provide a much more complete picture of life in ancient oceans than the normal fossil record

Figure B (a) *Marrella splendens*—one of the amazing fossils from the Burgess Shale and (b) its 3-D reconstruction viewed from the right side.

Source: Sinclair Stammers/Science Photo Library (a) and © Royal Belgian Institute of Natural Sciences (b)

of skeletal remains. This more comprehensive record allows the ecology of fossil communities to be investigated. Evidence of different feeding strategies (e.g. predator, herbivore, detritivore) links animals in food chains which are interconnected in a food web. As more ancient food webs are reconstructed it will be possible to explore how they differ in other settings and how they are impacted by extinction events through geological time. Soft-bodied fossils provide a unique source of evidence of many different aspects of the history of life on Earth.

❓ Pause for thought

Why was it so important that the shapes of the soft-bodied fossils from the Burgess Shale were visualized accurately? What is the danger of including a structure which is actually an artefact of the fossilization process?

The chances of an organism becoming fossilized are extremely small. Most organisms die, and their bodies are destroyed by decomposing organisms. Fossilized organisms have somehow escaped some or all of this process. The fossil record is, inevitably, an incomplete story of the development of life on Earth. Nonetheless, it is important for the following reasons:

- The fossil record provides evidence for key events in the story of the development of life on Earth. It is our only evidence for groups of animals that became extinct, such as the dinosaurs and the curious creatures in the seas of the Burgess Shale.

- The fossil record does give an indication of the biodiversity of life present in the rocks at that moment in time, which allows us to glimpse primitive ecosystems. It also allows us to monitor the change in biodiversity over time.

- The fossil record provides an indication of the passage of time, independent of modern biochemical and genomic studies. It provides a way of validating other types of studies.

- The fossil record remains the core evidence base for macroevolution, the large-scale evolution of families, phyla, and kingdoms.

What stories can fossils tell?

Scientists have used the evidence from fossils to help them unravel the evolution of organisms from bacteria to humans (see Chapters 2, 7, and 8). The story below and the bigger picture 3.1 will give you insights into just some of the evolutionary trails we have uncovered in this way.

How the dinosaurs got their coloured feathers

Simple feathers were not unusual in dinosaurs. Some scientists think that the skins of most dinosaurs were covered with 'proto-feathers' and that they acted as insulation, as hair does for mammals. The fossilized remains of a dinosaur called *Kulindadromeus zabaikalicus*, was found in the Kulinda region of southeasten Siberia. It was a small herbivore with a long tail (Figure 3.5a) which lived in the mid-late Jurassic period, which ended 145 mya. It was covered in simple feathers, each with a central shaft but without the barbules that hold modern bird feather barbs together on the vane (Figure 3.5b). In this respect, they resembled the feathers of Silkie hens (Figure 3.5c).

By the early Cretaceous period, some theropod dinosaurs had feathers resembling those of birds. The feathers of *Sinosauropteryx*, found in rocks in China, had melanosomes, suggesting that the feathers were coloured. *Sinosauropteryx* had dark coloured stripes on its tail and probably was coloured in chestnut to reddish brown tones, as shown in Figure 3.6.

Some later theropods, like *Archaeopteryx*, developed wings and could fly. These are the ancestors of birds. Natural selection ensures that adaptations that increase the survival of organisms will be passed onto future generations. Over time, these adaptations might gain new functions. The proto-feathers that kept *Kulindadromeus* warm could, through natural selection, have evolved the additional function of signalling. Coloured dinosaurs might have a considerable advantage in the sexual

Figure 3.5 (a) *Kulindadromeus zabaikalicus,* (b) a modern bird feather, and (c) a Silkie hen feather.

(a)

20 cm

(b)

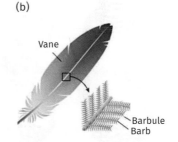

Vane

Barbule
Barb

(c)

Source: (a) By Tomopteryx— Own work, CC BY-SA 4.0 (b) Nadtytok/Shutterstock (c) PLOS.org

Figure 3.6 Scientists have good evidence that *Sinosauropteryx* had coloured feathers— another stage in evolution.

Source: Bee_acg/Shutterstock

selection of mates. Later dinosaurs, with developed forearms, might gain an advantage from gliding. This might, in turn, develop into flight. Feathers did not evolve to enable birds to fly: they evolved because they conferred a selective advantage in survival. Flying emerged only when the necessary anatomical components were already in place.

The evidence from anatomy

Anatomy is the study of the structure of living organisms, the many ways that parts are organized into bodies. Comparing the anatomy of different groups of organisms gives us clues to the evolutionary relationships between them—see Figure 3.7. Here are some examples of what we can learn.

How the vertebrates became so dominant

Similarities and differences in the structure of the heart and circulatory systems of vertebrates suggest clear trends in evolution, shown in Figure 3.7.

Fish are the oldest and most primitive vertebrate group, emerging about 530 mya, during the Cambrian Explosion (Chapter 2). The heart consists of a single receiving chamber (atrium) and a single pumping chamber (ventricle). Blood flows through the heart once on its journey through the gills and round the body. This is called a single circulation.

Figure 3.7 The anatomy of the heart and circulatory systems of different groups of vertebrates.

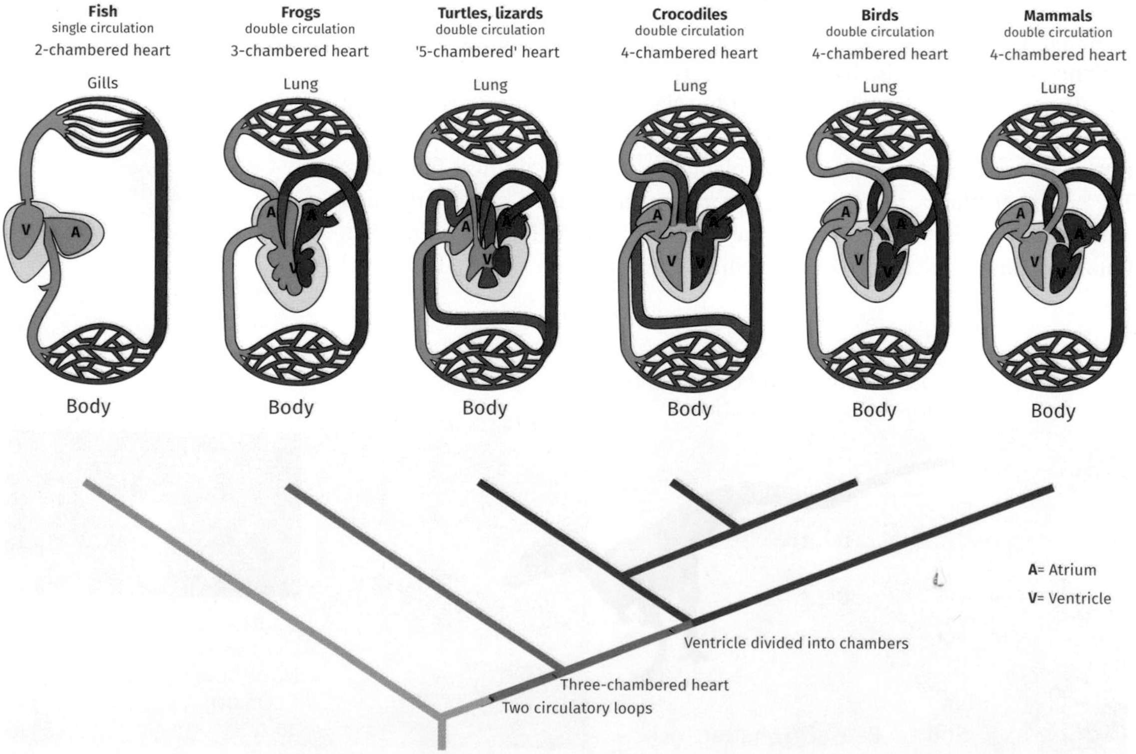

A= Atrium
V= Ventricle

Ventricle divided into chambers

Three-chambered heart
Two circulatory loops

The formation of separate circulatory mechanisms for the gas exchange systems (the pulmonary circulation) and the rest of the body (systemic circulation) seems to have occurred in advanced fish such as the lungfish, which evolved about 380 mya. The lungfish is now thought to be very similar to early land animals. It has pairs of primitive lungs, used for gas exchange. It also has gills, but, in most living forms, these are no longer used for gas exchange. The lungfish has a heart with a separate left and right atrium. The pulmonary vein carries oxygenated blood from the lungs to the left atrium. The ventricle then pumps the blood to the rest of the body in the systemic circulation. Since the blood passes through the heart twice on its journey round the body, it is called a double circulation.

Although this form of the heart evolved in lungfish, it is also the pattern found in amphibians, such as frogs and toads. This suggests that amphibians and modern lungfish shared a common ancestor about 370 mya.

Returning the blood to the heart after oxygenation to be pumped on around the body is a significant advance. The hydrostatic pressure of the blood in the systemic circulation in amphibians is much higher than in fish. However, any gains in efficiency are limited because the oxygenated and deoxygenated blood mix in a single ventricle, so the blood that is transported to the body has a reduced oxygen content.

An evolutionary trend seen in the reptiles is to separate the left and right ventricles by a muscular septum; in primitive reptiles, such as turtles and lizards, the separation is incomplete and there is a gap that allows mixing of the blood between chambers. In crocodiles, this hole is completely sealed, allowing full separation of oxygenated and deoxygenated blood. This is why crocodiles are capable of growing so large and swimming so strongly (see Figure 3.8).

We might predict that, based on their size and activity, dinosaurs' hearts would have fully closed septa, like crocodiles. However, without evidence, it is impossible to test this prediction. In 2000, scientists discovered a fossil dinosaur, *Thescelosaurus*, with an intact stone heart. Computerized imaging techniques suggested this was a four-chambered heart with a single aorta. This was a major finding, because it also implied the dinosaur might have been warm-blooded, a feature normally associated with birds and mammals. It caused great excitement amongst scientists. Disappointingly, a later analysis using advanced techniques

Figure 3.8 The differences between the hearts of lizards and crocodiles cannot be seen from the outside, but crocodiles have made a definite step up the evolutionary ladder. © Anthony Short

suggested that the heart was an artefact, formed by sand grains that had been cemented into the shape of a heart, with no 'chemical signal consistent with a biological origin'. The hunt for a real dinosaur heart goes on.

How the giraffe avoids a headache

Mammals, with their efficient four-chambered heart and double circulation, have successfully diversified to live in all parts of the Earth. Even though they share the same basic body plan, mammals show variation in their hearts, which have become adapted for survival in their diverse environments. These adaptations are often in several organ systems, as seen in the giraffe.

The giraffe has a unique set of selection pressures on its circulatory system. Standing about 5.7 metres tall, the heart has to pump blood, against gravity, up to the head and brain.

The heart of the giraffe is comparable with many mammals, with a mass of about 11kg, about 0.9% of its body mass. It has the standard mammalian four-chambered structure. What makes the heart of the giraffe unique is the thickness of the left ventricle wall and the small volume of the ventricle cavity. This pumps (relatively) small volumes of blood at twice the hydrostatic pressure of hearts of other mammals. The blood has the momentum to flow upwards to the head and brain. The arteries in the neck have wide lumens and significant amounts of muscle and elastic fibres to withstand the high levels of pressure generated by the left ventricle and keep the blood moving upwards.

Figure 3.9 The giraffe has evolved a special system which makes it possible to put its head down to drink without passing out.

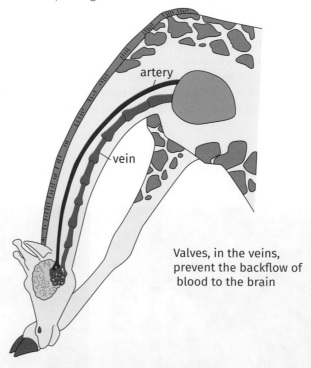

artery

vein

Valves, in the veins, prevent the backflow of blood to the brain

Source: The BioManual

But what happens when the giraffe lowers its neck to drink? How does it avoid its brain being crushed by the mass of the blood flowing downwards? There is an extensive network of small arteries surrounding the brain that are rich in elastic fibres. These expand to accommodate the blood flowing from the heart and reduce its pressure. The arteries prevent too much blood reaching the brain.

The veins down the neck, through which the blood travels back to the heart, contain many valves, which are open when the head is raised, but are closed when the neck is lowered at the drinking pool, preventing the flow of blood back into the giraffe's brain (Figure 3.9).

How the giraffe got its long neck has fascinated biologists for hundreds of years. It is a question we shall revisit and try to answer in Chapter 4.

The evidence from biochemistry
The molecular clock is ticking

The science of biology depends upon the development of tools and techniques that allow new insights into the natural world. Hooke used the newly-invented microscope to open a window into the world of cells and revolutionized our understanding of living organisms. Similarly, the development of automated, computerized tools for biochemical analysis has enabled systematic surveys of the macromolecules found on Earth to be completed. Proteins and nucleic acids, especially DNA, are routinely studied in this way.

The arrangement and order of amino acids in a protein is determined by the sequence of sets of three nucleotides on part of a molecule of DNA. The ordering is determined by the genetic code, which is the same in all living organisms. The genetic code is something that all organisms have inherited from LUCA.

Mutations in the DNA code may lead to amino acid substitutions in proteins.

Comparing the number of DNA base or amino acid substitutions across different groups of organisms can provide evidence for possible evolutionary pathways.

Groups with a small number of differences shared a common ancestor more recently than those with a larger number of differences.

Furthermore, if we assume that these substitutions occur at a constant rate, independent of environmental conditions, then the number of differences between groups is proportional to the time since they diverged from the common ancestor.

These are molecular clocks, which are important sources of evidence for evolution. Comparisons of the macromolecules of different organisms have led to fundamental revision of how we think about the evolution of life on Earth.

Life emerges as domains

We saw in Chapter 1 that Charles Darwin foresaw the kingdoms of nature as being 'genealogical trees', each linked back to a single common ancestor. The examples discussed so far are descriptions of how individual branches form and separate on the tree of life. How can we step back and study the tree itself?

Carl Woese and his colleagues achieved this in 1977 by surveying variation in a small RNA molecule, containing about 1500 nucleotides, that forms part of the structure of ribosomes. Ribosomes are cytoplasmic organelles where amino acids are assembled into proteins. Their analysis has fundamentally changed how we think about the organization of life on Earth. Figure 3.10 shows his new classification.

Before Woese, the existing classification system was built around five fundamental kingdoms, animals, plants, fungi, protists, and bacteria. Differences in the organization of cells suggested that bacteria were thought to be significantly different to the other kingdoms. Bacteria were classified as prokaryotic and the remaining kingdoms were eukaryotic.

Woese took this further by creating a new top-level group called the 'domain'. The prokaryotic bacteria are separated into two domains, primitive Archaea and more advanced Bacteria. The third domain, the Eucarya, contains the established eukaryotic kingdoms as well as groups, such as the slime moulds, that do not fit comfortably into any of the established groups. This classification shifts the spotlight away from the 'great kingdoms', re-emphasizes the importance of the microbiological world, and shows how the major domains are Darwin's 'genealogical trees', rooted in the common descent from LUCA. Without comparative biochemistry, this could never have happened.

The evidence from ecological genetics

For many biologists, evolution takes place in the outside world, not in a biochemistry laboratory. Darwin proposed that evolution occurred when heritable variation interacted with the environment. In such a situation, ecological genetics would the natural way to study evolution. Each of the

Figure 3.10 Woese's diagram of the three domains.

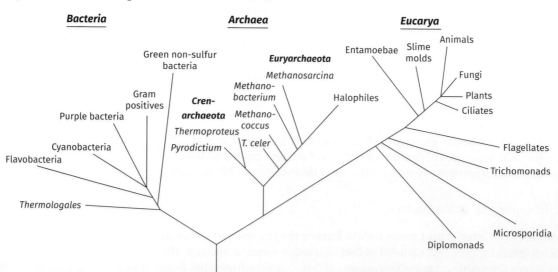

species discussed in this book could be investigated in such a way. In this section we will give a detailed account of one of the more remarkable species on Earth.

How the monarch butterfly learned to navigate

The monarch butterfly is an iconic organism in the United States. Each autumn some populations undertake a long-distance migration from the North Eastern states to a few overwintering grounds in Mexico. Then, in the spring, they journey back to their home states. In some years, hundreds of millions of butterflies overwinter in Mexico (see Figure 3.11).

The migration cycles of monarchs are synchronized to the calendar, with the first butterflies arriving in Mexico around November 1st, coinciding with the local Mexican celebrations for the Day of the Dead. They leave around March 15th. The migrations involve butterflies migrating considerable distances, of up to 5000 km and it is claimed that they return to the same trees in Mexico each year to hibernate. Yet the butterflies that arrive each winter are not the same ones that departed the previous spring.

Figure 3.11 (a) Monarch butterflies clustering together; (b) monarch migration route south; and (c) monarch migration route north.

(b) Migration south

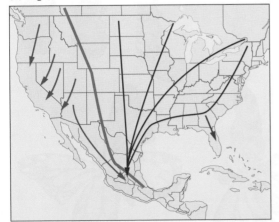

(c) Journey north

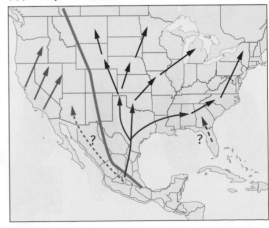

Source: (a) Noradoa/Shutterstock

Figure 3.12 Painted lady butterflies are UK migrants, sometimes visiting as part of their huge migration across Europe to Africa and back. © Anthony Short

Long-range, seasonal migration is quite common among insects (see Figure 3.12). Annual migrations suggest that this trait is heritable, and that the genome can be expressed differently in the different seasons. Darwin said that heritable variation is the prerequisite for evolution.

The monarch butterfly has a typical Lepidopteran life cycle (Figure 3.13):

egg>five larval (caterpillar) stages >pupa>adult

The caterpillars feed almost exclusively on milkweed plants. In spring, the butterfly sets off on its migration from its overwintering grounds in Mexico back to the Northern United States, undergoing up to five cycles of sexual reproduction during the journey. The butterfly that completes the journey is not the same butterfly that left Mexico. Then, in the autumn,

Figure 3.13 The life cycle of a monarch butterfly.

individuals undertake the reverse journey south to the overwintering grounds. Many of these butterflies complete the journey in a single generation. The development of these butterflies is arrested to prevent their reproduction and increase their longevity. This is caused by a reduction in the biosynthesis of juvenile hormone (JH).

How do the monarch butterflies find their way back to the same overwintering trees in Mexico? One explanation is an internal map, but this would have to be encoded in the genome, the only structure inherited from generation to generation. Alternatively, monarchs have, within their antennae, a sun compass which navigates using the position of the sun in the sky. This is regulated and corrected by a circadian clock under tight genetic control. There is a second magnetic navigation system used on cloudy days.

The evidence suggests that monarchs navigate using compasses rather than internal maps, but the map hypothesis cannot yet be rejected due to a lack of experimental research. Even now, scientists are not sure why and how millions of butterflies converge on a few sites in Mexico occupying less than 1 km^2.

Whilst the butterflies are overwintering in Mexico, the sun compass reverses itself, in response to at least three weeks of cold conditions experienced in the overwintering sites. The reversal in the compass enables the butterflies to undertake the homeward journey.

What is clear is that the annual migrations have been going on for about a million years, and that the butterflies have co-evolved with the milkweed plant, the primary food resource for the larvae. The northwards migration of the butterflies is co-ordinated with the emergence of the new milkweed plants, with the butterflies following the plants. Milkweed plants contain cardenolide poisons. Over time the monarchs have accumulated genetic changes which not only give them immunity to the poisons but also enable the caterpillars to accumulate the poisons. They have evolved bright yellow, black, and white striped markings—typical warning coloration. Predators see the colours and avoid eating the caterpillars.

Genes shedding light on evolution

Comparison of the DNA of monarch butterfly populations across the world is shedding light on the natural selection pressures acting on the monarch genome.

About 2% of the butterfly genome has been selected for migration, involving about 536 protein-coding regions (genes). These are active in the formation of the different stages of the life cycle, and the development of the nervous system and flight muscles.

Unravelling the mystery of the migration of the monarch butterfly has involved detailed, reductive, experimental work in many fields of biology. It is only when these independent pieces of research are integrated together that a rounded picture of the migrations emerge.

Now, however, the annual monarch migration, seen for a million years, is experiencing a devastating decline and there is fear that the phenomenon may disappear entirely. Climate change and anthropogenic habitat loss are affecting the milkweed plants on which the butterflies depend, highlighting how sensitive these sophisticated evolved systems are to disruption in the environment. Fortunately, research empowers people to try to conserve the butterfly by growing milkweed plants in their gardens and by promoting their welfare. There is still time.

The evidence from genomics

Comparison of genomic differences between migrating and non-migrating butterflies was an important tool for unravelling the mystery of the monarch butterflies. Genomic evidence is being widely used to tell us something about the formation of species. The genome describes the total DNA found in the organism. Whilst most DNA is found in the nucleus of eukaryotic organisms, small amounts are found in mitochondria, and in chloroplasts in plants. Mitochondria are only inherited from the mother, in the egg cytoplasm, and the DNA can be used to trace inheritance along the female line.

Correspondingly, in humans, DNA on the Y chromosome is only inherited from the father, and can trace inheritance down the male line. For example, it is claimed that Genghis Khan, the founder of the Mongol Empire in Asia (see Figure 3.14), had an unusual sequence of nucleotides on the Y chromosome called haplotype C-M217. It is claimed that about 1 in 200 men in the world carry this sequence, and are therefore descended from the emperor overlord. Khan died about 750 years ago, 30 generations ago, so such estimates are possible. If you go back far enough, everyone is related to everyone else.

We can now sequence all of the individual DNA nucleotides relatively cheaply and rapidly (see **Genomics** in this series). Many nucleotides are the same in all individuals. Such nucleotides are said to be fixed in the population— they are often part of regions that carry the genetic code for proteins (genes), and are fixed as a result of natural selection. Surprisingly, such protein-coding regions occupy only a small part of the total genome (1% in humans). The non-coding regions are not subject to natural selection and can accumulate variation in bases, such that a G (guanine) might be replaced with a T (thymine).

Nucleotides that show variation in the bases are called single nucleotide polymorphisms (SNPs). There about 10 million SNPs in the human genome, out of a total of 3 billion pairs of nucleotides distributed across 23 pairs of chromosomes. Some SNPs form sequences of up to six or more base pairs that may be repeated, often many times. Such sequences are called microsatellite DNA. In non-protein-coding regions of DNA, microsatellites have no known function, but they are useful markers for genetic fingerprinting and ancestral research.

We now have a reliable, cost-effective technique for collecting evidence of the evolutionary relationships between organisms and reading the story of natural selection in the past (see Case study 3.1).

'Evo-devo' and the power of the developing embryo

Developmental biology has been able to capitalize on the emergence of genomics. Not only can the role of DNA in the development of embryos be investigated, but the evolution of DNA through time can also be inferred. The science of evolutionary developmental biology (affectionally called 'evo-devo') is becoming one of the major contributors to the revolution in evolutionary thinking that is already taking place (see Chapter 4 and another book in this series **Developmental biology: embryos, evolution, and ageing**). We will illustrate its power by considering one of its earliest success stories, the role of **homeotic genes** in embryonic development.

In the 1980s scientists analysed a set of eight genes in the fruit fly, *Drosophila*. In the fly embryo, five genes controlled the organization of the thorax and three genes organized the abdomen. Within each of these

Figure 3.14 Genghis Khan: based on genomic evidence, you or someone you meet in your life will be one of his male descendents.

homeotic genes, scientists found 180 nucleotide base pairs which were identical. They coded for a 60 amino acid sequence enabling the protein produced to bind to DNA. These are transcription factors, proteins that control gene transcription and, therefore, protein synthesis.

The order of the homeotic genes is important. Scientists found that rearranging the order caused bizarre effects: flies were formed with legs emerging from their head, where antennae should normally form. These genes are clearly big players in development: the homeotic proteins are the master switches that activate networks of other genes during development.

Homeotic genes are found throughout the animal kingdoms, and it seems they evolved at least 558 mya, well before the Cambrian Explosion. The key nucleotide sequences are identical in all animals. Homeotic genes from *Drosophila* can control development in mice (see Figure 3.15).

Homeotic genes can explain the similarity in animal forms across the natural world, but flies, mice, and humans show significant differences as well as similarities. Can homeotic genes explain these too? Although the protein-coding nucleotide sequences (genes) are conserved throughout the animal kingdom, the ways in which the genes are expressed in different tissues is not. The expression of genes is usually controlled by regulatory sequences of DNA upstream from the protein-coding regions. A gene could have several of these regions active in different tissues. Sometimes the regulatory sequences can be some way from the gene, possibly on different chromosomes. Mutations in the regulatory sequences can occur and be used by natural selection to create a range of different effects. Regulatory gene systems, of which homeotic genes are but one example, are toolkits in the process of natural selection.

The evolution of the melanic form of the peppered moth, described in Chapter 1, shows that, in the right environmental conditions, a new dominant allele can spread through a population in less than 100 years. The exact nature of the melanic mutation is not known, but it is thought to have been in one of the 'master switches' that control development, rather than in a gene that produces an enzyme that makes the pigment.

Figure 3.15 Homeotic genes control embryo development in a wide range of species.

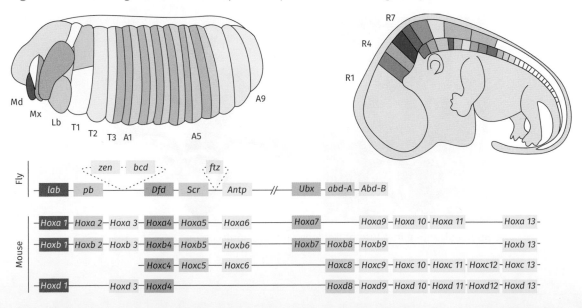

The bigger picture 3.1
The mysterious origins of flowers

Our understanding of the evolution of modern species is being built up by an interweaving of different types of evidence, illustrated here by the story of flowering plants.

In a letter to Joseph Hooker in July 1879, Charles Darwin wrote 'The rapid development ... of all the higher plants within recent geological times is an abominable mystery.' The evolution of flowers has remained a mystery until very recently. Now advances in techniques for studying fossilized flowers and also plant genomics are starting to unlock the secrets of the origin of flowers.

Darwin's problem was that the fossils available to him were mostly from the late Cretaceous period (from 100–66 mya), *after* there had been a significant period of evolution in which most of the modern-day groups of flowering plants emerged. This is called the **adaptive radiation** of flowering plants. To Darwin, it seemed as if all the different families of flowering plants evolved at the same time.

We now know that magnolias, eucalypts, and primitive grasses all emerged during the Early Cretaceous period (approximately 125 mya). A recent discovery of a fossil flower dating back to the mid-Jurassic period (174 mya) has pushed this revised timescale back still further. This flower, shown in Figure A, is *Nanjinganthus dendrostyla*—amazingly, a contemporary of some of the dinosaurs.

Most flowers consist of male and female reproductive organs (stamens and carpels), surrounded by petals and sepals. The avocado plant (*Persea americana*) is a modern descendent of the primitive magnoliid plants that emerged at the start of the Cretaceous period. Compare it directly with *Arabidopsis thaliana*, which evolved later, in Figure B. The *Persea* flowers are simpler than those of *Arabadopsis* and there is no distinction between petals and sepals. They are called tepals in Figure B.

Figure A A fossilized flower of *Nanjinganthus dendrostyla*, and a computer-generated model of the flower as it might have looked.

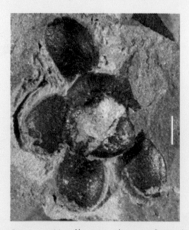

Source: Nanjing Institute of Geology and Palaeontology

Figure B *Persea americana* and *Arabidopsis thaliana*, with relevant areas of their genomes.

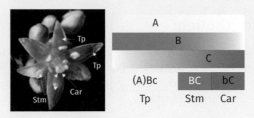

Persea americana

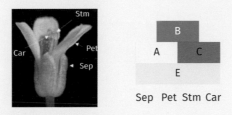

Arabidopsis thaliana,

*Tp = tepals, Sep = sepals, Pet = petals
Stm = stamens, Car = carpels*

Source: Chanderbali, A.S., et al. (2009). Transcriptional signatures of ancient floral developmental genetics in avocado (*Persea americana*; Lauraceae). *Proceedings of the National Academy of Sciences* 106 (22) 8929-8934; DOI:10.1073/pnas.0811476106

Studies of the parts of the genome active during flower development (the **transcriptome**) suggest that there are differences between *Persea* and *Arabidopsis*, which might help to explain the evolution of flowers. Flower development involves the production of many different proteins, so it is not surprising that it is controlled by networks of interacting genes—probably about 5000 in each type of plant. The networks of genes are activated by a smaller number of transcription factors.

The production of transcription factors is controlled by specific regions of the genome, some of which are active in both types of flower (A, B, C, in Figure B). *Arabidopsis* also has a region of the genome that is not found in *Persea* that produces a different transcription factor (E). The differences in the development of these types of flower can be explained by the way that these transcription factors work together.

This case study illustrates how natural selection can, over millions of years, cause real changes to the genome. This is most likely caused by duplication of critical parts of the genome, which then allows mutations in DNA bases changes to accumulate in the copy, allowing the original to retain its usual function. Gradually, over millions of years of natural selection, the base sequence of the copy becomes so different to the original that it develops a different function.

 Pause for thought

At what levels of biological organization is natural selection acting in the evolution of flowers? Is it acting on the phenotypes of the flowering plants, or on the alleles of the genes or on the genome? This is an important idea, which we will discuss further in Chapter 4.

 Chapter Summary

- The evidence for evolution comes from a wide variety of different fields of biology, including fossils, comparative anatomy, biochemistry, population genetics, ecological genetics, genomics, and developmental biology.
- Fossils are an important source of evidence, from Mary Anning's discoveries to modern finds of dinosaurs with feathers. A range of techniques have been developed to reconstruct the three-dimensional forms of organisms.
- Comparison of the anatomy of organ systems in different groups of organisms can help to reconstruct major trends in evolution.
- Mutations are changes to the base sequence of DNA, which may have effects on proteins produced by the organism. Mutations occur at random in the genome; the accumulation of mutations is used as a molecular clock to estimate the rate of evolution.
- Comparison of differences in RNA led Woese to propose that life evolved as three domains (Archaea, Bacteria, and Eucarya).
- Studying the evolution of an organism involves combining evidence from ecology, physiology, and genetics.
- Genomics is a powerful new source that allows DNA base sequences to be compared in a range of living and fossil organisms.
- The impact of evolution of the development of organisms provides powerful evidence of the conservation of DNA sequences that regulate development across millions of years of evolutionary history.

Further Reading

Li, Q.-G., et al. (2010) Plumage color patterns of an extinct dinosaur. *Science* 327, 1369– 1372.

Derek Briggs and his team's spectacular late Jurassic example of finding evidence of coloured dinosaur feathers.

Dawkins, R. and Wong, Y. (2004). *The Ancestor's Tale: A Pilgrimage to the Dawn of Life.* **Boston, MA: Houghton Mifflin.**

Beautifully told stories about the evolution of living organisms.

Gould, S.J. (1989). *Wonderful Life: Burgess Shale and the Nature of History.* **New York: W. W. Norton & Co.**

Highly readable, if controversial, account of the Burgess Shale fossils (see the next chapter).

Carroll, S.B. (2005). *Endless Forms Most Beautiful: The New Science of Evo Devo and the Making of the Animal Kingdom.* **New York: W. W. Norton & Co.**

The book on evolution for the genomic era. Utterly absorbing and fascinating from start to finish.

https://ucmp.berkeley.edu/cambrian/burgess.html

Accessible site on the Burgess Shale fossils

https://evolution.berkeley.edu/evolibrary/article/evodevo_01

Accessible site on evo-devo (evolutionary developmental biology)

http://news.bbc.co.uk/earth/hi/earth_news/newsid_8143000/8143095.stm

Accessible site on the evolution of flowers

https://www.thoughtco.com/evolution-of-the-human-heart-1224781

Accessible site on the evolution of the heart

https://www.worldwildlife.org/species/monarch-butterfly

Accessible site on monarch butterflies

 Discussion Questions

3.1 Some people argue that because evolution cannot be investigated by controlled experiments, it cannot be regarded as a scientific theory. What do you think?

3.2 Scientists think that you share about 355 genes with the Last Common Universal Ancestor that lived about 4 billion years ago. Suggest the processes these genes are most likely to affect.

3.3 Recently, in Mexico, there have been attempts to destroy the trees in which the monarch butterflies overwinter. Suggest what is the likely impact of actions like these on the survival of the monarch butterflies.

4 THE EVOLUTION OF THE THEORY: CHANGING VIEWS

You know you have arrived when people take your ideas and develop them into a way of thinking, putting *-ism* after your name. It happened to the Buddha, to Karl Marx, and even the former UK prime minister Margaret Thatcher. It also happened to Charles Darwin, although only after his death, and after several new discoveries were gradually incorporated into the theory. From 1895 onwards this revised theory of evolution was called neo-Darwinism.

A significant reboot to the theory, Evolution 2.0 if you like, was launched in 1942, under the name of the '**Modern Synthesis**'. The Modern Synthesis is more than a single theory, it is a set of tools for thinking that lets us build a model of the evolving living world. Our understanding of evolution continues to grow, and some scientists are calling for a further extension to the Modern Synthesis (Evolution 3.0).

This chapter traces the development of these tools for thinking by looking at the ideas behind the theory of evolution and speculates what Evolution 3.0 might look like.

Figure 4.1 Our theories of evolution seek to understand all of the biodiversity we see, in ecosystems from the harsh climate of Iceland (a) to a tropical rain forest (b). © Anthony Short

From Darwin to the Modern Synthesis

We saw in Chapter 1 that Darwin was not the only person thinking about the problems of the origin of species in the 19th century. Although it was the Darwin–Wallace model that became established, partly through the runaway success of Darwin's book, *On the Origin of Species*, other models of evolution were discarded along the way.

One of these was the thinking of Jean-Baptiste Lamarck on the inheritance of acquired characteristics. He developed the well-established idea that the environment could cause changes to the characteristics of an organism, and that these changed characteristics could then be passed on to future generations, making acquired characteristics a driver of evolutionary change. Contemporary discoveries in biology are leading some scientists to revisit these ideas—read on to discover more!

The Modern Synthesis emerged from the re-interpretation of the *Origin of Species* in the light of two major shifts in thinking: the re-discovery of the work of Gregor Mendel leading to the science of genetics, and the establishment of cell theory by August Weismann and the cell biologists.

Evolution meets genetics

The Weismann barrier is a very useful concept, but it does not explain how variation passes from generation to generation, a significant problem for Darwin's theory. Darwin, in common with many others, believed that variation was blended: the phenotype of an offspring being the average of its two parents. This is illustrated in Figure 4.2, where a new 'red' variation appears in an otherwise white population. It breeds with white and produces pink offspring. Repeated breeding with white in each subsequent generation will halve the intensity of the red colour.

Case study 4.1
August Weismann and the meiotic revolution

The German biologist August Weismann (Figure A) was a key thinker in the development of the Modern Synthesis. Ernst Mayr, himself one of the architects of the Modern Synthesis, called Weismann the 'second most notable evolutionary thinker of the 19th century'—second only to Charles Darwin!

Weismann helped to develop the technique of cytology, staining and slicing tissue so that it could be seen under a light microscope. The technique opened up the world of the cell, which underpins modern biology. The process of **mitosis**, by which most of the cells of the body divide, was named in 1882 by Walter Flemming. But in 1890 it was Weismann who recognized the significance of the two divisions seen in the process of **meiosis**, the form of cell division seen in the production of the gametes (Figure A).

Weismann's discoveries lead to a revolution in our thinking, called the Weismann barrier. The idea was to think about the **germ cells** (the cells that divide to produce gametes) as being wholly separate to the **somatic cells**, which make up the rest of the body. Only changes to the germ cells can be passed on to the next generation through inheritance, and so contribute to evolution. Changes to somatic cells might significantly affect the ability of an individual to survive, but they cannot be inherited by future generations.

Weismann's barrier is a major tool for thinking about evolution.

- Firstly, it excludes Darwin's pangenesis theory (see Chapter 1), which suggested that all cell types produce gemmules which travel to the reproductive organs to be incorporated into gametes for sexual reproduction.

- Secondly, it undermines any ideas that characteristics acquired by somatic cells can be inherited and thus play a role in evolution. It is little wonder that the Modern Synthesis utterly refuted any notions of Lamarckian inheritance. Natural selection could be the only major driver of evolution.

- Thirdly, it provides a cellular explanation for Darwin's speculation that all living organisms are descended from a single common ancestor (see Chapter 1). At fertilization, gametes fuse to form an embryo, which develops to form somatic and germ cells. The process continues from generation to generation, as it has since the emergence of the first cell.

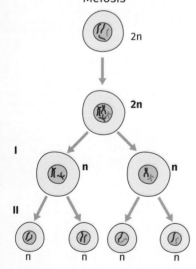

Figure A Mitosis and meiosis.

Mitosis

2n

2n

2n   2n

Meiosis

2n

2n

I

n   n

II

n   n   n   n

❓ Pause for thought

Suggest reasons why our understanding of evolution continues to develop.

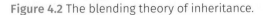

Figure 4.2 The blending theory of inheritance.

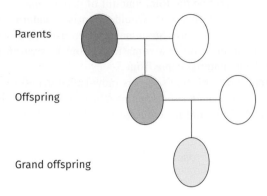

Parents

Offspring

Grand offspring

As a result, any new variation is unlikely to persist for many generations. By the 1880s, this was regarded as a fatal flaw in Darwin's theory which threatened, in Julian Huxley's phrase, to lead to the 'eclipse of Darwinism'.

The answer to this dilemma was proposed in 1865, but lay neglected for 35 years, until its rediscovery in 1900—Gregor Mendel's experiments on plant hybridization in peas. Mendel proposed that variation be considered as discrete 'factors', passed from generation to generation, predicting the patterns of observations found in the offspring of two parents that differed for a particular characteristic. This seemed to work well enough in pea plants, and the Mendelian champions, William Bateson and Reginald Punnett, showed that it seemed to work in other eukaryotic organisms too.

Mendel's idea, that variation could be reduced to factors which could be modelled mathematically, became a recurrent theme in evolution. At the heart of Mendel's model is the idea that the factors were present in pairs in the adult organism, but as single factors in the gametes. At fertilization, offspring were formed with two factors. Meiosis halves the chromosome number, and fertilization restores it.

The next stage was to reconcile the work of Weismann and his colleagues with the new thinking of Gregor Mendel. The big idea was that Mendel's factors had a physical existence as genetic material on chromosomes. Chromosomes were shown to exist in pairs in somatic cells, and singly in gametes. Fertilization brought together new combinations of chromosomes, from the father and the mother, to restore the adult number of chromosomes.

A new language of accountancy was needed: the term 'haploid' was used to describe the set of single chromosomes in a gamete, and 'diploid' for the set of pairs of chromosomes in a somatic cell. In 1920, Winkler proposed the term 'genome' to describe the total amount of genetic material in a haploid gamete. A somatic cell, therefore, would comprise a paternal and maternal genome. These days, the term 'genome' is often used to describe the total amount of genetic material in a somatic cell, but it is worth remembering where this genetic material came from.

The idea that linked Mendel's factors (now called genes) to chromosomes led to an explosion of interest in mapping genes onto chromosomes. This was led by Thomas Morgan and his colleagues in the United States, who worked

Figure 4.3 Some of the *Drosophila melanogaster* phenotypes that scientists have been able to link to specific genes.

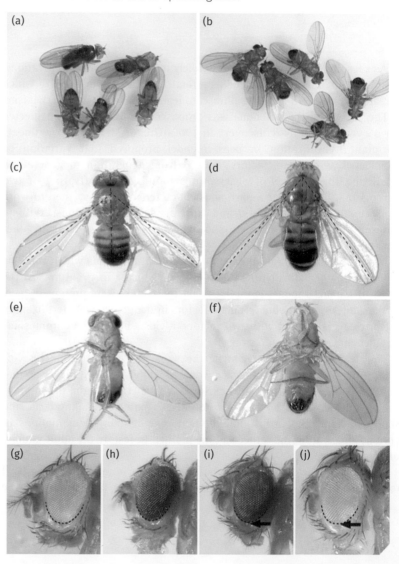

Source: Johnstone, K., Wells, R.E., Strutt, D., and Zeidler, M.P. (2013). Localised JAK/STAT Pathway Activation Is Required for Drosophila Wing Hinge Development. *PLoS ONE* 8(5): e65076. https://doi.org/10.1371/journal.pone.0065076

with a local fruit fly called *Drosophila melanogaster*. They systematically mapped the variations seen in the adult flies, such as eye colour and wing size, to discrete genes located on its four pairs of chromosomes (see Figure 4.3). It was painstaking, remarkable work, and led to the conception of genes being arranged along a chromosome like 'beads on a string'. This is a breathtakingly powerful metaphor—we can only speculate what Darwin would have made of it, when he was writing the *Origin of Species*.

Moving the model forwards

The scene is almost set for the unveiling of the Modern Synthesis, but there is one last question to answer—where does new genetic variation come from? If we reject the idea that variation produced by the environment can be inherited, then we are left with the alternative that all inherited variation can only arise through changes in the genetic material—in the genes. The work of Morgan's team showed, beyond doubt, that these changes must be located in chromosomes. Such changes are called mutations.

The problem was that most characteristics are not like Mendel's characteristics. Mendel chose 'unit characteristics' that existed in two contrasting forms, such as tall and short pea plants or yellow and green seeds. Each of the characteristics could be modelled by a single factor and (eventually) mapped to a particular location on a chromosome.

Most characteristics are not like that. Continuously varying characteristics (such as height) involve many different genetic factors interacting with the environment. They cannot be mapped to chromosomes using Morgan's methods and are studied using statistics. Mutations for these characteristics have small, gradual effects on the phenotype. These ideas affect how we see evolution. Small effects would lead to evolution proceeding in small, progressive steps (which Darwin favoured). Mutations with larger effects might cause evolution to proceed in discontinuous 'jumps', called 'saltatory' evolution.

This was a serious stand off and undermined the credibility of natural selection as a driving force in evolution. The resolution came from Ronald Fisher, a statistician. He took Mendel's ideas and reformulated them, so that they applied to continuously varying characteristics such as height. His approach, called biometrical genetics, provided a theoretical foundation for the emerging Modern Synthesis. Biometrical genetics also, incidentally, indirectly laid the foundation for the sciences of genomics and bioinformatics.

Fisher's book, *The Genetical Theory of Natural Selection*, was written in 1930, and it proposed that most mutations would reduce the biological fitness of an organism, although some could be neutral and have no effect. A very few would be beneficial. By fitness, Fisher meant the ability of an organism to survive and reproduce. In order to be inherited, these mutations would need to be in germ cells. Thus, evolution would proceed with a small, but regular, supply of beneficial gene mutations. He went further and proposed a geometrical model for fitness: where the potential fitness of a new mutation would be inversely proportional to its effect on the phenotype. Thus, beneficial mutations would be most likely to have only a small effect on the phenotype.

The question of whether evolution occurs in small progressive steps or saltatory jumps was left open. These days, most scientists accept that evolution can proceed in both ways, depending upon the time, the environmental conditions, and the species. The evolution of life during the Cambrian Explosion (Chapter 2) provides examples of saltatory evolution. The directional and stabilizing selection in the evolution of the Galapagos finches (Chapter 1) are examples of a more progressive, continuous change.

Natural selection is the main driver of evolutionary change, which is observed as changes in allele frequencies in populations. The Modern Synthesis is, primarily, concerned with gene pools (populations of genes) in populations of organisms.

By now, you might be wondering what remains of Charles Darwin's original idea. Actually, it remains alive and well at the heart of the Modern Synthesis. Darwin's ideas, which we outlined in Chapter 1, were reinterpreted by Mayr and are still the way that we approach understanding evolutionary changes in populations. Here we present them as series of generalized observations and deductions:

- Observation 1: All species have a great potential to increase in number.
- Observation 2: The numbers of most species remain approximately constant.
- Deduction 1: There is a struggle between individuals for existence. Many individuals die before they can reproduce.
- Observation 3: Much variation within populations is inherited. Organisms have inherited characteristics.
- Deduction 2: Some inherited characteristics adapt organisms to survive in their environments.
- Deduction 3: Natural selection will favour those individuals that have the most effective inherited characteristics.
- Deduction 4: Over the generations natural selection will gradually change the inherited characteristics in a population and may lead to the production of new species.

The Modern Synthesis (see Figure 4.4) was published in 1942 by Julian Huxley, the grandson of Thomas Henry Huxley, who championed Charles Darwin's cause all those years ago. It was, and remains, a cornerstone of evolutionary thinking.

Building on solid foundations

The Modern Synthesis fiercely rejects the idea that characteristics acquired during an organism's life could be inherited by its offspring, and thus play a role in evolution. This was largely as a result of Weismann's thinking about the separation of the somatic (body) and germline (reproductive) cells.

The dangers of rejecting evidence-based science

There were other reasons why the architects of the Modern Synthesis were keen to reject the inheritance of acquired characteristics. Soviet Russia, from the 1920s onwards, operated an agricultural policy named after its founder,

Figure 4.4 The Modern Synthesis of evolution.

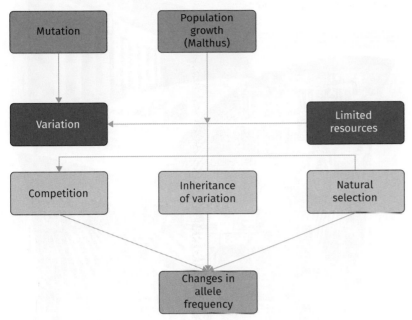

Trofim Lysenko, with the full support of the leader Joseph Stalin. Lysenkoism (as the policy was called) was a collection of pseudo-scientific practices that supported the Soviet ideal that society could be transformed by hard work and collective effort (Figure 4.5). Darwinism's natural selection was replaced by 'natural co-operation', Mendelism was rejected in favour of the inheritance of characteristics acquired through life. In people, that would mean hard work and a collective pulling together, in plants, a lifetime of growing as well as possible. So rye wheat would be transformed into the more valuable bread wheat if the workers tried hard enough and the plants grew well. ...

The policy was a disaster: agricultural productivity fell, leading to food shortages, and over 3000 biologists lost their jobs. Tragically, many were imprisoned, and some were executed. Genetics research in Russia was effectively stopped until after the death of Stalin in 1953. The impact on the people of Russia was even worse. Millions died of starvation, followed by millions more in China, where they followed Russia's lead. It is estimated that Lysenkoism lead to around 30 million deaths from lack of food. This is a terrible reminder of the potentially fatal consequences of using carefully selected scientific evidence to support a political ideology.

The Modern Synthesis firmly closed the door to the inheritance of acquired characteristics and there was no political will to reopen the door for the rest of the century.

The impact of the human genome

The publication of the Human Genome Project from 2000 onwards (Figure 4.6) caused shockwaves through the whole of biology, and the Modern Synthesis may not be immune. Evolutionary thinking may be

Figure 4.5 Lysenkoism—replacing natural selection by 'natural co-operation'—destroyed the lives of many biologists and led to millions of deaths from starvation. Source: C. and M. History Pictures/Alamy Stock Photo

Figure 4.6 The Human Genome Project changed the way we think about genetics and evolution forever.

undergoing a paradigm shift and an extension to the Modern Synthesis might be needed. There are certainly people arguing for this, as well as those arguing against. You may recognize a common theme here: evolutionary theory only develops after a lot of arguments!

The original genome project proposed that humans have 30,000 genes, (although this has been revised downwards)—fewer genes than the number of proteins in the human body. Thus, the idea of:

  gene> protein> characteristic

had to be reconsidered. Furthermore, the discovery that the information needed to code for the amino acids in a protein were not, in eukaryotic organisms, necessarily located as continuous blocks of DNA on the same chromosome began to undermine the neat idea of genes being 'beads' on a chromosome 'string'.

The picture emerging was more like the thinking of the Biometricians and Fisher: characteristics emerge from large networks of genes interacting with each other and with the environment. This is now the dominant model in the systems-based view of biology.

Biologists at the turn of the 21st century started looking for evidence to support this viewpoint, and rediscovered the work of Conrad Waddington. Waddington was an English developmental biologist, who learned to experiment with fruit flies in Morgan's laboratory. He viewed evolution from the perspective of the developing organism, and was critical of the way the Modern Synthesis reduced evolution to changes in allele frequencies in populations. It was not, he argued, a naturalist's view of evolution that Darwin would recognize. Waddington performed two experiments which suggested that genetic systems could interact with the environment to create novel characteristics, and that these could be inherited. Moreover, they could do this without violating the Weismann barrier—see Case study 4.2.

Case study 4.2
Waddington and Genetic Assimilation

In 1953, Conrad Waddington was studying the development of the wings of *Drosophila*. Wild type flies have cross veins on their wings (see Figure A). Waddington induced a phenotypic change in a line of **wild type** flies by giving the young fly pupae a strong environmental shock (40 °C for four hours). This affected the development of the flies, producing a range of phenotypes including some which lack the cross veins on their wings ('crossveinless' flies, Figure A).

The heat shock did not produce gene mutations which caused this change. It altered the normal environment of the flies, disrupting their development. The changes induced by the heat shock could not be passed on to the offspring of these flies.

About 40% of Waddington's flies showed the crossveinless phenotype. He selected them, mated them together, and reapplied the heat shock to the pupae. He repeated the breeding and selection programme for about 25 generations. Eventually, he produced flies that would always develop the crossveinless phenotype, even when the heat shock treatment was no longer applied.

Waddington had discovered a new phenomenon, which he called the 'genetic assimilation of an acquired character'. In other words, the characteristic that had been induced by the environment had been assimilated into the genomes of the flies.

Waddington explained genetic assimilation using a metaphor called the 'epigenetic landscape', a model that visualizes the interactions between the genome and the environment. This is shown in Figure B.

Figure A Normal and crossveinless *Drosophila* wings.

(a)

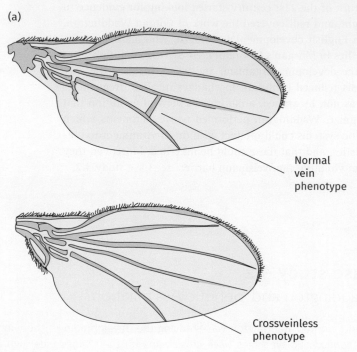

Normal
vein
phenotype

Crossveinless
phenotype

Figure B presents the epigenetic landscape as it is usually drawn, as a gently undulating valley sloping down towards the observer. A ball, representing the fly embryo developing inside a pupa, is presented with a series of alternative pathways of development, as represented by the channels. As the embryo progresses down one of the channels it will change and develop into its adult phenotype.

The deepest channel represents the normal 'wild type' developmental pathway. Another channel will represent the developmental pathway towards the crossveinless phenotype.

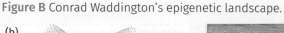

Figure B Conrad Waddington's epigenetic landscape.

(b)

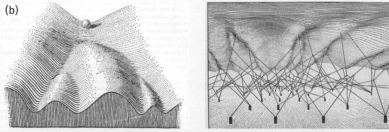

Source: Waddington, C.H. (1975). *The Evolution of an Evolutionist*. Ithaca, NY: Cornell University Press.

Pathways of development are shaped by the organism's genome, represented by the network of tent pegs underpinning the landscape in b of Figure B. This is a key point: it is not one or two genes working in isolation, it is the whole of the genome that works to shape development—here, Waddington was decades ahead of other contemporary thinkers.

The heat shock disturbs the normal equilibrium of the landscape, moving the ball so it no longer travels down the channel towards the wild type phenotype. Instead, it moves down a different pathway that leads to the crossveinless phenotypes.

Selection of the flies brings together combinations of alleles that subtly change the shape of the landscape, making it easier for the embryo to travel down the 'crossveinless' pathway, when provided with the environmental stimulus. Gradually, and this takes at least 20 generations of selection, the landscape is so changed that the 'crossveinless' pathway becomes the default channel down which the embryo develops. At this point, the environmentally induced effect becomes fully assimilated into the genome.

This is not quite the inheritance of acquired characteristics envisaged by Lamarck and rejected by the Modern Synthesis, because Waddington's model does not violate the Weismann barrier. All parts of the embryo—those destined to be somatic cells and germ cells—are equally exposed to the same environmental shock and both develop along their own epigenetic developmental pathways.

❓ Pause for thought

In what ways is the genetic assimilation proposed by Waddington similar to, and different from, the inheritance of acquired characteristics proposed by Lamarck?

Evolution 3.0

In 2005, Eva Jablonka and Marion Lamb wrote a book, *Evolution in Four Dimensions*, questioning the gene-centred modern way of thinking about evolution. Looking at systems of inheritance, Jablonka and Lamb suggested four different ways that characteristics could be inherited across the generations: genetic, epigenetic, behavioural, and symbolic. A rounded appreciation of evolution, they argued, could only come by thinking about how these systems interact with each other. The book is almost a manifesto for extending the Modern Synthesis into Evolution 3.0.

Much of this book so far has explored the significance of genetic inheritance in evolution. Inheritance can involve vertical gene transfer, from parents to offspring in sexual reproduction or horizontal gene transfer, directly between organisms, seen in the movement of plasmids between bacteria and now thought to have had a major role in the evolution of eukaryotic cells.

In this section we will briefly consider the other systems discussed by Jablonka and Lamb, because they are areas of active research, and may well be where major discoveries take place in the future.

Epigenetics

Waddington's research, discussed in the previous section, has matured into a discipline of its own called epigenetics. Epigenetic markers are chemicals that attach to DNA (eg the addition of methyl groups to the base cytosine) and the histone proteins that surround the DNA in chromosomes (eg acetyl groups attaching to lysine amino acids on the proteins). Some RNA molecules can also be markers. Whilst the markers do not affect the sequence of DNA bases in protein-coding regions, they do affect how these regions are expressed. The markers can change the shape of the proteins around the DNA (chromatin remodelling). It is rather like adding a volume control to a gene: the activity of the gene can be turned up, or down or switched off completely.

Epigenetic markers are passed on to daughter cells during mitosis, and, in plants and microbes, can be transmitted across the generations. Therefore, epigenetic markers have played an important role in the evolution of multicellular organisms.

There is some evidence of transmission of epigenetic markers across generations in mice and in humans, although this seems to be restricted to two generations. This is because epigenetic makers are removed during meiosis and the formation of gametes. This allows organisms to adapt their gene expression in response to current environmental pressures, rather than those of their parents or grandparents.

The ability to respond epigenetically seems to be under genetic control, and therefore the ability to respond epigenetically to changes in the environment is an adaptation of genomes to a changing world.

Closely allied to epigenetics are transposable elements (or transposons), small pieces of DNA which move to different places within a genome. Often, they duplicate before moving and so leave a copy of themselves in the original location. Depending on their new location, they can alter the expression of genes, create new mutations, or even reverse the effects of other mutations. Barbara McClintock discovered transposable elements in maize, which won her a Nobel Prize in 1983. Astonishingly, about 90% of the maize genome is made up of transposons. In humans it is about 44%, some of which have been introduced by retroviruses. The 'gene' that causes the wings of the peppered moth to turn black is actually a transposon disrupting a regulatory gene called *cortex*.

Behavioural systems

Behavioural systems are only now being recognized as powerful forces shaping evolution. Organisms can change their local environment to improve life for themselves in a process called niche construction. In doing so, they can affect the survival of other organisms. Beavers provide a clear example of this effect in action. Beavers are mammals that build dams (see Figure 4.7), creating lakes which alter the flow of rivers. This affects the entire ecosystem, dramatically changing the organisms found in an area and their interrelationships. Such changes in the environment resulting from the behaviour of one particular species can drive the evolution of many other species. Beavers are being introduced into some aquatic ecosystems because they increase biodiversity. Some might argue that any evolutionary developments driven by climate changes in response to anthropogenic global warming are examples of behavioural systems impacting evolution—the behaviour being that of human beings.

Symbolic systems

Symbolic systems of inheritance involve the transmission of information through signs and signals (see Figure 4.8). Flowers and the coloured feathers of dinosaurs are examples of structures that became adapted to transmit symbolic information. It is, perhaps, an alternative way of viewing sexual

Figure 4.7 A beaver building a dam—behaviour that can impact the evolution of many other species.

Source: Chase Dekker/Shutterstock

Figure 4.8 The spectacular throat pouch of this male anole lizard, and the stunning colour of the male Adonis blue are purely symbolic—they signal to females but have no practical use at all. © Anthony Short

selection as a driver of evolution. A number of organisms have developed communication and the use of symbols, from the complex constructions of bower birds to the exquisite but useless tails of the peacock. Human language and culture are symbolic systems that have been important in our own evolution—see Chapters 5 and 6.

Interactions and evolution

We increasingly recognize that all of these systems of inheritance interact as evolution takes place. This increases the ability of organisms to adapt to their environments—known as evolvability. The same genotype can produce different phenotypes depending on the environment. The water crowfoot (Figure 4.9) produces different types of leaf, depending upon whether their buds are in water or air. This phenotypic plasticity increases the ability of the water crowfoot plant to respond to the water levels in the stream.

This new thinking on evolution is more unpredictable and less deterministic than Darwinism, neo-Darwinism, or even the Modern Synthesis. Natural selection can act at the level of the gene, or the organism, or at the level of the population. Some would argue that it can act at the level of the biosphere (through climate change, for example). It can, as in the example of insects and flowering plants, be acting on two different species at the same time.

It can also act on several levels at the same time. This is called multilevel selection. For example, many organisms host populations of microbes within their guts, which can help the host organism adapt to its environment. The woodrat is able to digest creosote and tannins because of the microbial community in its gut. This ability is lost when the microbiota are killed by antibiotics. Natural selection for the ability to digest creosote and tannins is acting at the level of the microbiome and the rat. Multilevel selection can also be illustrated by asking one of the oldest questions in evolution, 'just how did the giraffe get its long neck?'

Figure 4.9 Water crowfoot with the leaves it forms when it grows in water—a different version develops when it grows on land.

Source: Martin Fowler/Shutterstock

The bigger picture 4.1
How the giraffe got its long neck

We saw in Chapter 3 some of the complex adaptations shown by giraffes that enable them to drink safely. A question which has always intrigued biologists is 'just how did the giraffe get its long neck?'

For Lamarck, the long neck developed to reach the leaves at the top of tall trees. In times of drought, he argued, the leaves at the top of trees would be eaten last, so the tallest giraffes would stretch their necks to reach them. A 'nervous fluid' would flow into its neck and make it longer. Its offspring would inherit the longer neck.

Darwin, writing in the *Origin of Species*, was in broad agreement:

> the individuals which were the highest browsers and were able during dearths [dry seasons] to reach even an inch or two above the others will often have been preserved (p. 178)

If Darwin was neutral on the idea of acquired characteristics, the Modern Synthesis was not. The neo-Darwinian explanation of the evolution was that:

The genes that promoted the growth of the neck arose as random mutations and conferred an increased chance of reproduction because of more efficient feeding, leading to healthier individuals. The alleles conferring this advantage would be passed onto the offspring and would increase in frequency in the giraffe population.

The giraffe has become the go-to example for high school students' understanding of the differences between the thinking of Lamarck and Darwin and appears in textbooks across the world. It seems almost churlish to ask if it is true.

The argument that a long neck offers a feeding advantage in times of drought has a number of serious flaws. Younger giraffes are smaller than adult giraffes, and male giraffes are up to 60 centimetres taller than the females—so the females and the young would suffer more in hard times. An evolutionary strategy that favours males at the expense of females and juveniles might be expected to cause unstable populations. Yet, there seems to be no evidence for this. Furthermore, there are other leaf-eating mammals in the African savannah. Why do they not show similar trends in evolution?

This is supported by evidence from watching wild giraffes. Simmons and Scheepers (1996) write that:

> during the dry season (when feeding competition should be most intense) giraffes generally feed from low shrubs, not tall trees; females spend over 50% of their time feeding with their necks horizontal; both sexes feed faster and most often with their necks bent.

If feeding competition is not the answer, what are the alternative explanations? Darwin did not think it possible that the neck could have developed separately to the other parts of the body, which seems plausible. The legs of a giraffe are very long, giving a long stride which means it can run at about 50 km per hour, giving protection against predatory lions—a definite advantage in evolutionary terms. But an animal with long legs finds it difficult to bend down, so a longer neck would facilitate drinking from a pool.

Fossil evidence suggests that the immediate ancestor of giraffes was *Bohlinia*, a mammal which probably evolved in Southern Central Europe about 8 mya (Figure A). *Bohlinia* migrated to temperate China and north India in response to climate change.

Once in East Africa, *Bohlinia* evolved into a number of closely related, but separate sub-species of the modern giraffe, *Giraffa*, in a habitat where food was plentiful.

Analysis of mitochondrial DNA sequences and non-coding DNA bases of modern giraffes suggest there are at least six separate DNA lineages, with little evidence of interbreeding between them. Thus, the population structure is more complex than at first appears. There must be mechanisms in place that ensure that female giraffes only breed with males of their own **haplotype**. This suggests that some element of sexual selection might be important in their evolution.

Males secure females for breeding by fighting rival males for dominance (Figure B). They hit their opponents with their heads, swinging their large necks to generate greater impetus. These fights can be violent: injury and death can occur. Larger-necked males are the dominant victors, and their reward is to mate with the receptive females.

Male giraffe neck vertebrae and skulls are larger and more armoured than those of females (which do not fight). Furthermore, the necks of males continue to grow throughout their lives, a characteristic of sexual selection.

The pattern of markings on the giraffe is known to be highly distinctive within the haplotypes, and under high levels of genetic control. This, combined with the large neck, would make a dominant male distinctively recognizable to the females in its territory, so the neck also plays a symbolic role in maintaining the integrity of the population.

The present consensus is that the 'necks for food' hypothesis is less strong than the 'necks for sex' hypothesis, although both effects probably happen simultaneously in the giraffe populations.

Figure A *Bohlinia*, a relatively recent ancestor of the giraffes.

Source: Alain Bénéteau

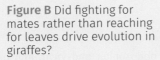

Figure B Did fighting for mates rather than reaching for leaves drive evolution in giraffes?

Source: Villiers Steyn/ Shutterstock

This case study shows the value of combining evidence from different fields of research to get a fuller picture of evolution. Natural selection is acting on the cells, tissues, and organs of the necks of the giraffes, on their skin pigmentation systems, on their genomes, on their group behaviour, on their mating behaviour, and on their response to predators.

❓ Pause for thought

Waddington argued that natural selection acted on organisms. Suggest why this includes all of the separate 'systems' described so far in this chapter.

Why does understanding evolution matter?

The remaining chapters of this book consider the remarkable rise of our own species. This is a good moment to pause and ask why evolution matters in society and is still the focus of debate, argument, and controversy. One reason why the theory of evolution is important is because science exists as a system of thought within human culture, alongside religions and philosophy. Evolution is not the only way of placing our own existence in a wider context. All religions have their own account of the origin of humans, from Adam and Eve in the garden of Eden for the Abrahamic faiths through to the Dreamtime of the Aboriginal peoples of Australia. Does the evolutionary version of events explained by the science community complement these other narratives, or does it replace them? This was such a controversial issue to Darwin that he excluded any talk of it from his book, *On the Origin of Species by Means of Natural Selection*—perhaps fearing that he was controversial enough already. He did discuss it at great length in a later book,

The Descent of Man (1871). The controversy did not end with the widespread acceptance of Darwin's ideas by scientists. The thinking of William Paley on Natural Theology allows us to explore this controversy further.

Consider watchmakers …

William Paley was an English philosopher and vicar, and in 1802 he published *Natural Theology: Evidences of the Existence and Attributes of the Deity*. In part of the book he carries out a thought experiment: walking in the countryside, he finds a stone. How, he asks, did the stone get here? He decides that, in the absence of any further knowledge, it had always been there.

Further on, he finds a watch on the ground (see Figure 4.10). This, he says, is different. The watch exists because it was designed and built by a watchmaker. Its structure is so complex it could not have formed on its own. If this is the case with a watch, Paley argues, then the natural world, which is so much more complex, must also exist because of an external intelligent designer.

The idea of an external designer of the living world was not new. The Greek philosopher Plato made a similar argument in about 400 BCE. He put forward the thesis that the universe is highly structured and organized, and its complexity suggests it has been designed.

For Paley and Plato, the very complexity of nature is powerful evidence for the existence of a designer. For most people in the UK and Europe in Victorian times, that designer was the Christian God. By explaining the complexity of nature in terms of natural processes and events, Darwin was seen to be undermining people's belief in this God.

This theme was picked up by Professor Richard Dawkins in his book, *The Blind Watchmaker* (1986).

> *"Natural selection is the blind watchmaker, blind because it does not see ahead, does not plan consequences, has no purpose in view. Yet the living results of natural selection overwhelmingly impress us with the appearance of design as if by a master watchmaker, impress us with the illusion of design and planning."*

Figure 4.10 William Paley's watchmaker analogy was based on an old-fashioned clockwork watch like this one.

Dawkins goes further, saying that natural selection dispenses with the need for God's existence. This is a controversial claim, certainly not supported by everyone. Michael Ruse, a philosopher of biology, argues that evolution 'does not necessarily' disprove the existence of God. Believing in a God remains, as it has always been, an act of faith and faith, by its nature, does not need scientific proof.

If natural selection is 'blind', with no forward planning involved, then where does this leave evolution? In this book, we have described evolution in terms of a series of processes, summarized in Figure 4.4. Evolution emerges when all of these processes work together and results in the organism best adapted for a particular set of circumstances at a particular point in time.

Many of us celebrate our connectedness with the rest of the living world through the extraordinary process of evolution, which has taken place on planet Earth, perhaps uniquely in the universe (Figure 4.11). For others, Paley's argument of design resulting in complexity remains powerful, and this book may provide evidence of something beyond the physical. Professor Michael Reiss, an evolutionist and an Anglican priest, has described these two positions as complementary world views. The evolutionary story of human origins complements, but does not necessarily replace, those accounts from other religions or philosophies.

Figure 4.11 The sheer variety, beauty, and number of living organisms is breathtaking. People have tried to explain it, in many different ways, for as long as records exist. © Anthony Short

Steven Jay Gould has argued that science and religion occupy different domains of thinking that do not overlap. He calls these Nonoverlapping Magisteria (NOMA). Science is associated with facts, whilst religions are associated with human purposes, meanings, and values. On the other hand, there are a number of people who have contributed to the evidence presented in this book who have professed a religious faith, including Father Gregor Mendel, Theodore Dobzhansky (an architect of the Modern Synthesis), Harry Whittington, and Simon Conway Morris.

If natural selection does not 'plan ahead', then random chance plays a significant role in evolution. It is a matter of chance that the Earth was struck by a comet about 66 million years ago, causing the sudden mass extinction of three quarters of the plant and animal species on the planet, including the non-avian dinosaurs. Steven Jay Gould, writing about the Cambrian Explosion seen in the Burgess Shale fossils, said that if we could rewind the tape of life, we might find ourselves living with very different organisms, because of all of the chance events that have shaped evolution. Simon Conway Morris, a biologist who worked on the Burgess Shale fossils, fundamentally disagrees. He argues that the tape of life can be run as many times as we like and, in principle, intelligence will emerge. On our planet, we see it in molluscs (octopus) and mammals (human beings, among others).

Conway Morris is describing the principle of 'convergence' where the same characteristic evolves independently in distantly related groups. The eye has evolved independently at different times in several diverse groups (such as insects, molluscs, and mammals). His argument is that sooner or later, if the conditions needed for life to exist are there, the processes involved in evolution will lead to the emergence of organisms with intelligence and consciousness, on Earth and elsewhere in the Universe.

If natural selection does not 'plan ahead' then we must be careful about the language we use when describing evolution. Evolution is not a purposeful 'goal-directed' system, although it sometimes seems like it—and people often write carelessly about it. For example, you might read that some dinosaurs evolved feathers in order to learn to fly, so that they could evolve into birds. This use of language suggests purpose and choice. It is called teleology, and is inappropriate and inaccurate when talking about evolution. The paragraph below tells the same story but is written to avoid teleological thinking. The way such things are presented has a huge impact on how people understand the world around them.

Feathers arose as a result of a chance mutation in some dinosaurs, which survived and had offspring with feathers. Gradually, over time, feathered dinosaurs gained a behavioural advantage because they could recognize each other easily. The feathers became used for signalling to potential mates, increasing reproductive success. Through natural selection, these feathers became larger and more effective signals. As a side effect, they helped the dinosaurs to stay warm, increasing their selective advantage. Fossil evidence shows that small predatory dinosaurs leapt into the air to catch their insect prey. Those with the larger feathers gained an advantage, because they could leap further and for longer. The parts of their genome promoting the growth of the forelimbs and feathers were shaped by directional natural selection and over time, wings with larger feathers formed. This led to the gradual emergence of animals that could fly.

The French biologist François Jacob used the word 'bricolage' to describe evolution. There is no exact English equivalent for this word—evolution acts like a 'tinkerer', a builder who does the best he can, using the materials he has at hand. It implies improvisation. The finished goods may have imperfections imposed by the limitations of the materials used, but they are functional. In evolutionary terms, the raw materials are the genomes, parts of which may be redeployed in new gene regulatory networks, or parts of genes, which are assembled to construct genes with new functions.

Other evolutionists have used different metaphors. To Waddington, evolution is not like an engineer building a machine from a fully designed blueprint. Natural selection is more like an artist, standing with paint and brushes at a canvas, allowing the painting to emerge by trial and error. Remarkably, this painting has somehow included you and me.

≋ Chapter Summary

- The theory of evolution has changed and developed since Darwin's *Origin of Species* in 1859. It underwent a significant revision with the Modern Synthesis in 1942 and continues to develop as new scientific discoveries are made.
- August Weismann's work on the separation of body (somatic) cells and reproductive (germ) cells by a barrier suggested that only genetic changes (mutations) in germ cells can be passed onto future generations in heredity.
- Gregor Mendel's ideas that genetic variation could be caused by discrete 'factors' that were inherited replaced other models of inheritance.
- Morgan, and other geneticists, located these genetic factors on chromosomes, and the genetic material was eventually identified as DNA. Gene mutations were defined as changes to the genetic code in DNA or in the organization of DNA on chromosomes.
- Most characteristics vary continuously in populations because a large number of factors (polygenes) interact with each other and with the environment. Individually, each polygene has a small effect on the phenotype.
- Modern Synthesis defines an evolutionary change in the frequency of alleles in a population, either under the influence of natural selection or by chance effects in populations of small size (drift). It combines contributions from genetics, cell biology, population genetics, and biochemistry into a model for evolution in populations.
- The spread of an allele though a population can be modelled mathematically.

- The inheritance of acquired characteristics, developed by Jean-Baptiste Lamarck, was tacitly accepted by Darwin, but firmly rejected in the Modern Synthesis. Recent research into the interactions of genomes and the environment shows they can be influenced by natural selection.
- There are calls for an extension to the Modern Synthesis of evolution because of the awareness of the importance of additional forms of inheritance, such as epigenetic inheritance, transposable genetic elements, behavioural inheritance, and symbolic inheritance.
- Natural selection is now thought to operate at different levels of organization within an organism (allele, organ system, or organism) or within the ecosystem.
- Understanding evolution is as much a philosophical or religious question as it is a scientific one. Each domain of knowledge has a different contribution to make to answering the question for each individual.

Further Reading

Huxley, J.S. (2009). *Evolution: The Modern Synthesis, the Definitive Edition*. Cambridge, MA: MIT Press.
The authoritative account of the Modern Synthesis.

Dawkins, R. (1986). *The Blind Watchmaker*. New York: W. W. Norton & Co.
Dawkins provides a dazzling account of the power of natural selection and an argument for the gene-centred view of evolution.

Jablonka, E., and Lamb, M.J. (2005). *Evolution in Four Dimensions*. Cambridge, MA: Bradford Books.
A look at the limitations of a 'gene-centric' view of the Modern Synthesis, using contemporary examples of evolution.

Conway Morris, S. (2003). *Life's Solution: Inevitable Humans in a Lonely Universe*. Cambridge: Cambridge University Press.
A provocative response to Steven Jay Gould's ideas in Wonderful Life.

Noble, D. (2016). *Dance to the Tune of Life: Biological Relativity*. Cambridge: Cambridge University Press.
Life as a network of interacting systems—this book is strong on the distinction between Darwinism and Lamarckism.

Discussion Questions

4.1 Have the discoveries in biology since the publication of the *Origin of Species* strengthened or weakened the theory of evolution?

4.2 Why is evolution called a 'theory' and not a 'law'—and suggest why this might cause problems for many non-scientists?

4.3 We think that all living organisms are descended from a single common ancestor (LUCA) that lived about 4 billion years ago. Review the evidence for this possibility. How does this influence the way we think about the place of our species in the living world?

5 HUMAN EVOLUTION: WHERE DO WE COME FROM AND HOW DID WE GET HERE?

Wherever we live, whatever we do, whatever the colour of our skin or the language we speak, we are all members of the same species—*Homo sapiens* (see Figure 5.1). Genome analysis shows that we share almost all of our DNA— the differences between us are very small indeed. We're also very similar to our closest living relatives—but somewhere along our evolutionary journey, we moved on to a different path.

In the next two chapters you will explore your own evolution. Where do we come from? Why do we walk on two legs instead of four, like the rest of the great apes? Why have our brains grown larger and larger over time, and our behaviour more and more complex?

We will start in this chapter with the living Primates and Hominids, which gives all scientists the basis for understanding our fossil ancestors.

The evidence we have is often fragile—the fossil record charting our development over time is, at best, incomplete and at worst, missing entirely. But the skill and imagination of many scientists, along with ever-more sophisticated ways of analysing the evidence we have, are giving us a clearer picture back through the mists of time.

Figure 5.1 All modern humans belong to the same species, a species that left Africa about 100,000 years ago.

Source: Rawpixel.com

Where do we come from?

To understand our own evolutionary history, we need to understand the factors which drove the evolution of the order of animals to which we, and our closest relatives, belong. As we have seen in previous chapters, the evolutionary process can only act on morphologies (shapes, colours, and structures) that already exist. The forces of evolution—gene flow, mutation, genetic drift, and natural selection—only act to mould a species through time, so we are constrained by the ancestral animals from which we descend.

For example, we stand upright, and walk bipedally (on two legs). The body structures needed to do this were not created de novo (starting afresh). They evolved from what was already present in our four-legged ancestors. Through time, we developed changes in the spine, hips, knees, and feet that enabled us to stand upright. But anyone who suffers from leg or back pain can tell you that this would not be the best way to create a bipedal animal! Bipedalism has painful consequences because it is not a new, perfect solution—it is a cobbled-together, best-we-can-do-with-what-we-already-had adaptation of an animal that already existed.

Figure 5.2 Although they all look different, primates share a suite of features in common: five fingers and toes, nails (not claws), emphasis on vision, larger brains, and generalized rather than specialized dentition.

Source: Iakov Filimonov/Shutterstock

Consequently, to understand our own evolution we must first examine our order, the Primates, and specifically the Great Apes, to which we are so closely related both anatomically and genetically (see Figure 5.2). This gives us information both about **how** we evolved, and the selective pressures driving these adaptations.

There are many ways of showing the relationships between ourselves and our fellow primates. Table 5.1 gives just one of them, with the taxonomic levels at the top, along with how the Primate order is organized within those groups. Scientists originally thought that humans separated from the great apes at the family level. Modern DNA evidence suggests we are much more closely related than that. We are now grouped with the African great apes at both the family and subfamily levels, separated only at the tribe level.

Although we often look to our closest relatives, the great apes (chimpanzees, gorillas, and orangutans), as well as the lesser apes (gibbons and siamangs), to see where we have come from, many of our adaptations can be traced to our earliest ancestors, and are therefore shared with all our relatives in the order Primates. As a group, we are conservative. We retain skeletal features widely found in the mammalian lineage, as opposed to being derived. What does this mean?

Consider the foreleg of a horse (see Figure 5.3). You can see that there has been a reduction in the number of bones compared to the bones in the forelimbs of other mammals, such as a human being. Horses' limbs are highly derived, because they have been selected for running speed. An elongated limb which moves forwards and backwards in a restricted plane allows horses to move very rapidly. But there is a price to pay for that speed. The specialization for running has led to a loss of digits, and so a loss of flexibility in how the forelimb can be used. It is *only* useful for running and standing. Primates, including Hominines, have been selected for

Table 5.1 Chart showing the relationships between us and our closest relatives. Notice that the family ends in 'dae' and tribe ends in 'ini'. Humans used to be the only members of the Hominidae, but we now recognize we are a lot closer to the great apes and are kept in the same family and subfamily.

Order	Suborder	Infraorder	Superfamily	Family	Subfamily	Tribe	Common term
P R I M A T E S	*Prosimil*						Loris Lemur Tarsier
	Anthropoidea	*Platyrrhini*					New World monkey
		Catarrhini	*Cercopithecoidea*				Old World monkey
			Hominaidea	*Hylobatidae*			Gibbon
				Hominidae	*Ponginae*		Orang
					Homininae	*Gorillini*	Gorilla
						Ponini	Chimp and Bonobo
						Hominini	Human

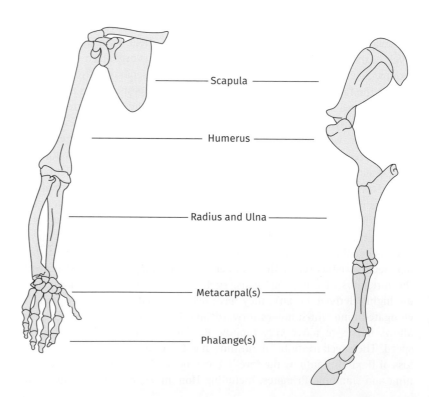

Figure 5.3 Retained and derived characters—the humerus, radius, ulna, metacarpals, and phalanges can be identified in both horses and humans, but they have changed dramatically in shape, size, and number.

Scapula

Humerus

Radius and Ulna

Metacarpal(s)

Phalange(s)

morphologies which improve our ability to grasp and manipulate objects, and we have retained all our digits. We may run slower than a horse, but we can do a lot more with our hands!

The characteristics that define primates as an order can be described by the advantages they bring to the group. These primate characteristics and their consequences are discussed below.

- **Sight over smell:** Primates emphasize vision, and de-emphasize the sense of smell. Many mammals have large, sensitive, wet noses with a very strong sense of smell, and black and white vision. Primates have small, dry noses and forwardly directed eyes, most with over-lapping fields of vision. These forward-facing eyes give the animal depth perception (see Figure 5.4). This is a hugely important ability when leaping from branch to branch in trees! Colour vision is also present in all the higher primates. Dogs and cats rely heavily on their sense of smell. Primates, however, are very visually oriented.

This emphasis on vision and de-emphasis on smell has both morphological and behavioural consequences. Primates are colourful in appearance and use visual signals to communicate. The food that most primates search out, ripe fruit for example, is found through colour. If you are walking your dog, you may be enjoying the scenery or looking for a restaurant, but your dog will be sniffing at everything to find out what other animals have been through the area before him. He will be smelling food or mating possibilities. Vision will be providing a secondary input for him. The change to forwardly directed eyes reflects an evolutionary shift, placing visual information over the sense of smell, and this has implications for the structure of the primate face. Our faces are flat, and our eyes are large and forwardly directed. Most primates have faces that we can identify with, and facial expressions become very important in the primate lineage.

- **Five fingers, five toes:** Primates generally retain the primitive mammalian condition of five fingers and toes and have replaced claws

Figure 5.4 Look at the eye positions and the noses of the dog and the squirrel monkey (a primate)— very different evolutionary forces have produced very different sets of sense organs. © Anthony Short

with nails. This allows us to better investigate our environment with our hands, to make things, and to carry things. Primates can grasp using fingers and thumb. This is possible because we have replaced claws with nails. The opposable thumb, with its great degree of flexibility, also has very important evolutionary consequences—see later.

Using our hands and feet to investigate our environment is something all primates do. We pick up objects, we squeeze them, we peel fruit, and ultimately, we make tools. While many primates make simple tools, humans are the ultimate toolmakers. It is this ability that has allowed us to modify and control our environment, and has led to our ultimate adaptation—culture.

Finally, as we will see below, our ability to grasp means that primates tend to carry their infants about with them and those infants can watch and learn. They are not left in nests, but cling to their parents as the parents engage in all kinds of activities (see Figure 5.5). This learning period is essential to train the larger primate brain.

- **Teeth for all diets:** Primates retain a generalized dentition, with incisors, canines, premolars, and molars, which gives a variety of dietary options, from plant material to other animals. Again, this is a retention of the generalized mammalian condition. Herbivores like horses have a specialized and reduced dentition. They have large incisors to cut grass and large molars to mash it. Not much else is needed. Carnivores, on the other hand, need specialized teeth to kill their prey and tear meat from the bones (see Figure 5.6), but it is more difficult for them to chew plant material. As a group, primates keep all their options open!

Figure 5.5 The ability to cling means that baby apes don't stay in a nest. They travel with their parent, constantly watching and learning.

Source: Photoholgic/Unsplash

Figure 5.6 Notice the difference in both the shapes and the number of teeth in the different skulls. Teeth are metabolically expensive to make, so you only make those you can use.

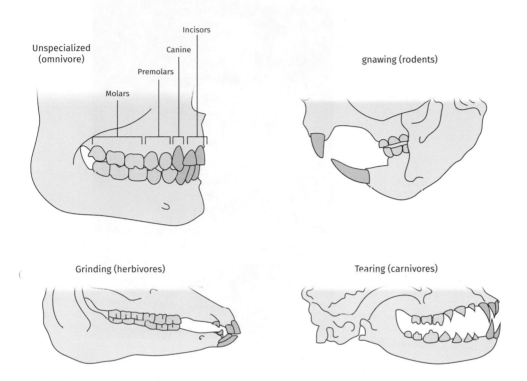

As primate teeth are generalized, our diets are also generalized. Some smaller primates eat primarily insects, and have pointier teeth to make that possible. Many primates eat fruit, leaves, and nuts and have basined, mashing molars, while a few specialize on leafy vegetation and have shearing cusps. But in most cases, the incisors, canines, premolars, and molars remain and are only comparatively slightly modified to adapt to specific diets.

- **Bigger brains = more flexibility:** Primates show increased brain size when their body size is taken into account, compared to other mammals. This increase in brain size leads to an increase in behavioural plasticity, so primates rely less on instinctive behaviour than other mammals—they can modify their actions and learn different behaviours. These larger brains need more training, and primates show a corresponding lengthening of the periods of development, called ontogeny. The time spent as an infant, juvenile, adolescent, and young adult are all longer in primates than in any other mammals. A dog is ready to reproduce at 6 months, but a female chimp needs 6 years to reach sexual maturity. During that time, she will follow other females about, attempting to help hold, carry, and play with other infants. This learning period is needed so that, when she has her

Figure 5.7 Human babies take years to become independent of their parents. © Anthony Short

own first baby, she will know what to do and is more likely to rear it successfully. Female chimps deprived of this time within a family group do not know what to do with an infant, and often neglect and fail to rear their own babies.

Humans have carried this brain size enlargement to extremes. Our brains are so big that we give birth early in development, before the skull is fully formed, so that the large-brained infant can exit the birth canal without crushing its skull or getting stuck. As a result, we produce extremely defenceless and helpless infants—humans have the longest ontogenic periods of all primates. Most primates can cling from birth, but baby humans cannot even cling to their mothers until they are several months old (see Figure 5.7).

The Last Common Ancestor

When we look for evolutionary relationships between the primates, we use the same approaches whether we are looking at the relationship between monkeys, apes, and their last common ancestor, or the relationship between the great apes and man, and their last common ancestor.

Today we can determine relationships among modern animals by looking at their DNA. Genetic similarities indicate evolutionary closeness. But when fossils are introduced into the picture, with very few exceptions, there is no DNA—although Neanderthals are a notable exception (see Chapter 6). To determine if a fossil is more closely related to one species than another, we must look at the morphology, searching for similarities. In hominin fossils we ask questions such as:

- How big is the brain?
- How much has the skeleton been altered?
- What changes do we see in the legs, hands, and feet?
- What changes do we see in the skull, face, jaws, and teeth?

Since we know from genetics that chimpanzees share 96–99% similarity with humans (depending on what is measured and who is doing the measuring), we can say from modern data that they are our closest relative. But how long ago did we separate from the chimpanzee line? Which fossil species belong on the branch to humans, and which do not? To find the Last Common Ancestor of man and chimpanzees (referred to as LCA) anthropologists have come up with a suite of characteristics that relate to our locomotion (bipedalism), intelligence (brain size, behavioural plasticity, culture), and dietary changes (face and dentition changes, tool use). We can then extrapolate back in time, comparing fossils and noting similarities and differences. We need to bear in mind that both chimpanzees and humans have been evolving since we last shared a common ancestor, about 5 mya.

The tribe Hominini—humans and our direct ancestors

The Hominini are the tribe that contains humans and human ancestors, including several side branches that went extinct millions of years ago. We know from genetics that the chimp is our closest relative, so we are looking to identify those species that evolved since the LCA—in other words, since we last shared an ancestor, about 5–6 mya.

To work this out, we need to identify the major differences between ourselves and the rest of the great apes. These lie in the fact that we walk on two legs, we carry things, we make things—and we have a big brain that gives us ideas about what to make. We also cook our food and have a diet which is different to that of the apes. Our jaws and teeth are different and so is our face. We use language to communicate. Consequently, when we look at fossils that might be related to us, we look for similarities and differences that relate to locomotion, brain size and indications of behaviour patterns linked to a larger brain, and our teeth, jaws, and face, which give us information about our diet.

The next section will define the characters that make us human: bipedalism, facial morphology and dentition, and brain expansion (along with the evidence of the behavioural changes that occurred as our brains got bigger).

Bipedalism

We are unique among primates, and in fact most mammals, in that we habitually walk upright on two legs. There must have been strong selection pressures for us to do this, because it requires major changes to the skeleton.

Although there are many theories as to why we evolved a bipedal lifestyle, scientists are still not sure what caused this dramatic change. The most accepted idea at the moment is that being bipedal frees our hands, it allows us to carry things for longer distances (such as food), and it also allows us to make and use tools (see Figure 5.8).

The changes that enable us to walk effortlessly on two legs can be seen in the skeleton in the skull, spine, hips, knees, and feet. When we find fossils representing any of these parts, we can be fairly confident in identifying the animal as bipedal. The fossil record shows us that these characters start developing about 6 mya and we see fully bipedal ancestors at about 3 mya.

We start to find stone tools at about 2.5 mya, but we need to bear in mind that we are relying on the presence of *stone tools*. Tools made out of bark, wood, leaves, and other similar materials would not be preserved over millions of years. It is very probable that the earliest tools were made of materials that would not be preserved in the archaeological record, so we cannot confidently link the two at present.

Figure 5.8 Clockwise from top left: Chimpanzees cannot stand upright for very long or go any distance on two legs—although this chimpanzee is certainly trying! Humans can carry very large loads and can walk great distances. The man making a stone tool will remove flakes from the stone in a particular way until he has achieved the tool he wants. Chimps use tools, but they are usually unmodified, or only modified to a very limited extent.

Source: a—E.D. Torial/Alamy Stock Photo, b—Branislavpudar/Shutterstock, c—Avalon.red/Alamy Stock Photo, d—Gorodenkoff/Shutterstock

Changes in the skull

If you are to walk any distance, it is best if your head balances comfortably on the top of your spine. It would be very tiring to use your neck muscles to hold your head erect when you were standing up! We can see the evidence of the neck muscles on any skull fragment that is from the back of the skull. On apes, who walk quadrupedally—using four legs—the muscles are very large and direct towards the back of the animal. In a biped, the muscles are smaller and direct downward. In addition, the foramen magnum—large hole—through which the spinal cord exits the skull and enters the spine, is placed towards the front, in the middle of the base of the skull. This balances the skull on top of the spine. In quadrupeds, the hole is towards the back of the skull. This little bit of anatomy is enough to distinguish whether a fossil belongs to a bipedal individual or not (see Figure 5.9).

Changes in the hips

The changes in the hips and the way they articulate with the femur are very dramatic in bipeds. In quadrupedal apes, the wide bone of the hip, called the ilium, is flat. It supports muscles that, when flexed, allow the leg to extend backwards. In bipedal humans the ilium curves around making a bowl shape, which supports the organs when we are upright. But most importantly, the outer surface of the bone creates a site for muscle attachments (gluteal muscles) that are on the side of the joint as well. This means we can extend our legs from the hip not only backwards but to the side (like a ballerina).

Why is this important? Because when we walk, although we are not aware of it, we are momentarily balancing our weight on each leg alternately. This

Figure 5.9 These diagrams show the position of the neck muscles and the foramen magnum in humans and chimps, and how they are arranged in humans to balance the skull on the upright body.

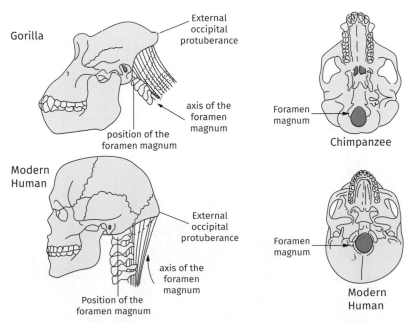

means that our entire body weight and centre of gravity shifts from leg to leg. The side muscles allow us to do this. You can feel this by pushing your fingers into the side of your leg where it meets the hip—and then just walk! You will feel these muscles contract alternately as you walk. Without this ability, you would fall over.

This seems such a small thing, but it is essential for bipedal walking for any distance. Consequently, fossil fragments of the ilium and hip region can often identify if we are looking at a hominin and may even tell us *how* bipedal they may have been (see Figure 5.10).

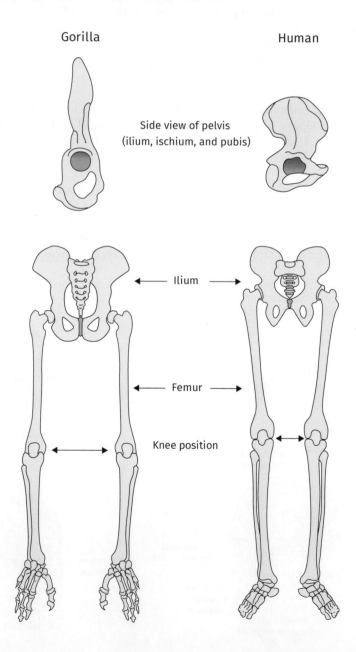

Figure 5.10 The ilium of the walking human flares around to the side, surrounding the joint between the hip and the femur. The gorilla ilium is flat and the muscles attach to the back of the joint. The lower skeletons (hips and legs) show the angling of the femur to bring the knees closer to our centre of gravity, and the thickening at the distal end.

Changes in the knee

Along with the hip changes in bipeds, we also see changes in the knee joint. We are, in fact, quite knock-kneed! Our femur runs from the side of the hip joint to the centre of our stance. This means that not only can we shift our weight easily, given the changes in the hip, but we don't have to make a very dramatic shift because the knees are close together. Again, you can demonstrate this by standing with your knees intentionally apart. Try to lift one leg and you will immediately fall over.

We can clearly see evidence of this evolutionary change in the knee joint. The lower part of the femur has built up extra bone to support the stresses on the outside surface of the bone. If you find a fossil with only the distal—far—end of the bone, it will be very obviously a biped if it has this feature (see Figure 5.10).

Changes in the foot

The feet show two important changes in the bipedal hominins. The first is that the big toe is pulled in or adducted to bring it in line with the other toes. This supports the entire foot and provides a push-off point with the big toe when walking. The second provides both transverse and longitudinal arches in the way the foot bones are articulated. The arches, supported by ligaments, serve to act as shock absorbers when the foot hits the ground, and also store energy to aid in the push off for the next step (see Figure 5.11).

When the foot evolved in this way, it became a very useful structure for running, walking, and standing, but fairly useless for manipulating objects.

Figure 5.11 These figures show the adducted big toe and the position of the transverse and longitudinal arches in the human foot compared to the gorilla foot, which lacks these evolutionary changes.

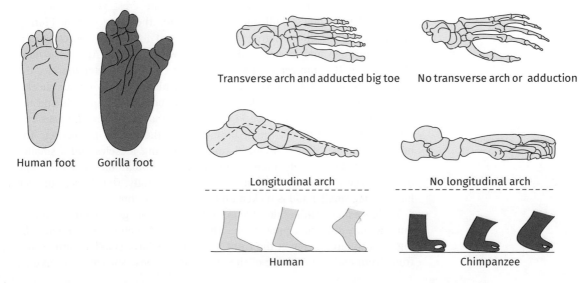

Human foot   Gorilla foot

Transverse arch and adducted big toe   No transverse arch or adduction

Longitudinal arch   No longitudinal arch

Human   Chimpanzee

While gorillas, chimps, and other primates use their feet in much the same way they use their hands, humans have lost that ability in the trade-off to bipedalism.

Facial morphology and dentition

Gorillas and chimpanzees are predominantly vegetarian in their diet. Chimps, in particular, are known to kill small mammals and eat meat, but this is an opportunistic special addition to their diet—it happens only when a good opportunity presents itself. Chimpanzees will band together and collectively kill a vulnerable monkey, for example, then share the meat—but the bulk of their diet is fruit. Gorillas eat plants, vines, grasses, and shrubs as well as fruit. The human diet shifted to include a much larger proportion of meat and, as we will see later, this shift had an impact on our brain size, behaviour, skull and face morphology, and dentition.

Uncooked vegetable foods are difficult to chew. To deal with their largely plant-based diet, gorillas and chimps have evolved large front teeth (to break off the vegetation and to bite through tough fruit rinds), and large, basined back teeth to mash it. They also have very large upper canines but this is to do with social structure rather than diet, and the canines sharpen themselves against a lower premolar. The chewing muscles are also large. When an animal has big teeth and big muscles, they need correspondingly big bony crests and shelves to provide surface for the muscle attachments. These factors account for many of the bone structure differences we see between the skulls of humans and apes. The crests and shelves are larger in the gorilla because the gorilla diet is much tougher, and male gorillas in particular need to eat massive amounts of food. Chimpanzees do not have the massive sagittal crests we see in gorillas.

Two large muscles—the temporalis and masseter—are attached to the sides of the skull and the jaw. The ape brain is small, so the skull has a large crest on the top of the skull for the attachment of the temporalis muscle—the sagittal crest. This is similar to the crest you can feel on the top of your dog's skull and it is there for the same reason. Flaring zygomatic bones allow a large space—temporal fossa—for the temporalis to pass through and attach to the lower jaw—mandible. Apes also have cheek bones that project for the attachment of the masseter muscle, the muscle you can feel if you clench your teeth while pressing on your cheeks. The jaws are correspondingly large to support both the temporalis and the masseter chewing muscles (see Figure 5.11). What we see in the apes is a large face grafted onto a small braincase, so the face is projecting. The hominin change of diet produced smaller chewing muscles and a smaller face, which is fitted onto a larger braincase. The human face, even in some of our earliest ancestors, does not project as strongly. Over time, our face has become smaller and is tucked under a large cranium.

We have small canines, but the large canines of the gorilla and chimp require a bony shelf to support them. This results in their dental arcade taking on a squared shape at the front. Our mouths have smaller canines and front teeth and so the shape of the arcade is rainbow-shaped or parabolic.

Figure 5.12 The bony crests and shelves for muscle attachment, and the size and shape of the teeth, which in turn affect the shape of the jaw, are major characteristics which help identify a human skull from an ape skull.

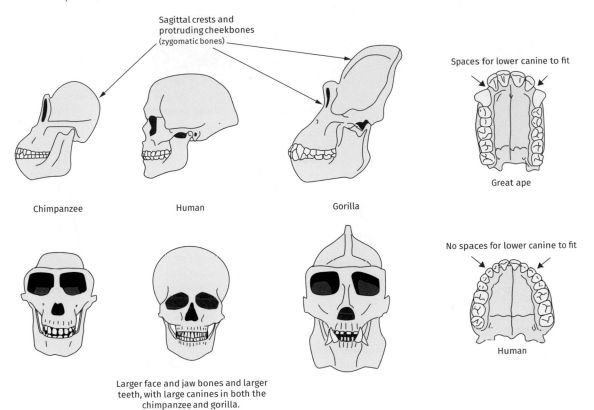

We also do not have to provide extra spaces for the large canines to fit when the mouth is closed, but apes do (see Figure 5.12).

As a result of all these evolutionary adaptations, there are clear differences between a human skull and the skulls of our nearest ape relatives. These are the differences scientists look for when fragments of fossilized skulls are found—and they help determine whether a fossil is hominin or ape in origin (see Figure 5.12).

Brain size

The characteristic that has the most impact on how we live our lives is our very large brain. Figure 5.13 shows how brains have got bigger, measured by cranial capacity because the volume of the cranium roughly equates to the size of the brain. Brains are metabolically costly to produce, maintain, and train, so there needs to be a very good reason for us to have this expensive organ. It is our brain that has allowed us to distance ourselves a little from the direct impact of natural selection, and enabled us to adapt

Figure 5.13 The great apes have a cranial capacity of around 400 cm³. For modern humans it is around 1400 cm³ on average. Interestingly there have been hominins with larger brains, such as Neanderthals at 1700 cm³.

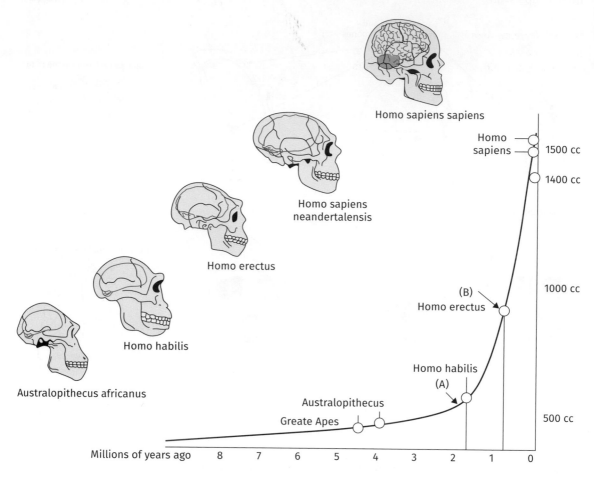

culturally in many different ways. This is not to say we are not evolving (we are), or that we are not subject to natural selection (we are), but we can, through culture, decrease the impact of selection in some cases.

For example, when we look at fossil people, we can see they usually have very strong and straight teeth. There was strong selection for good teeth because, in the absence of modern dental care, a tooth abscess could kill. Today, because our large brains have given us the ability to develop good dentistry, we see all kinds of variations in teeth. Many people spend a lot of time at the dentist to correct crooked and weak teeth, and as a result of this dental work they remain healthy and survive to pass on these dental problems to their children. This is not a source of concern; we handle many such problems through culture. Can you think of other ways in which we solve biological problems culturally?

Alongside the increase in brain size in fossils, we start to pick up archaeological evidence of cultural behaviours, which we deduce have developed as a result of the larger brains. This includes the use of fire, stone tools, and evidence of group hunting. We also see the first migration of early humans leaving Africa, where our earliest evolution took place, and spreading out all over the old world. *Homo erectus*, with a cranial capacity of around 1000cm^3, migrates as far east as China and as far west as the Middle East. Clearly an increase in brain size, along with the cultural adaptation it made possible, allowed our early ancestors to tackle problems presented all over the globe, and inhabit a wide variety of environments.

What triggered this rapid expansion of brain power? We don't know for sure but many scientists suspect it had to do with a shift in diet away from predominantly vegetables to more meat—and lots of it. *Homo erectus* was a hunter and followed migrating herds of game. He produced stone tools and hunted in groups. One suggestion by Leslie Aiello and Peter Wheeler was that as early humans became better at hunting, and hunted collaboratively, their diets contained much more meat, rich in protein and fat, which is much easier and more efficiently digested than plant material, and which is also higher in calories. This fed into a loop—the higher meat diet enabled early hominins to evolve a smaller gut, as less digestive system was needed to digest the higher quality diet. The gut, like the brain, uses a lot of energy, so a smaller gut and food of higher nutritional value made more resources available for the development of a larger brain (Figure 5.14).

In the next section, we will examine how what we know about our differences from the great apes helps us interpret the fossil record and recognize our ancestors. The characters and changes we have discovered so far are the same ones we will be looking at as we unravel our past. As explained earlier, the LCA of apes and humans will be difficult to identify. The LCA will retain characters seen in both modern apes and humans but not exclusively in either. If a fossil shows characters that are clearly derived on one line or the other, it could not be a common ancestor to both. We are actually looking for the earliest recognizable hominin, a fossil species that is part of our human lineage just after separation from the apes. To find our earliest

Figure 5.14 Aiello and Wheeler's flow diagram showing how higher protein (meat) might have led to increasing brain size.

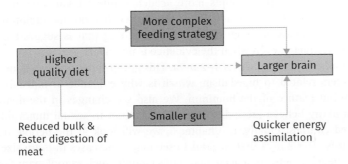

Source: The Wenner Gren Foundation for Anthropological Research

Figure 5.15 From the radiation of ape species that occurred in the Miocene, we can hypothesize that there is an LCA, but this individual will be hard to identify. The first hominin will be the species that shows the earliest evidence of at least one unique hominin characteristic.

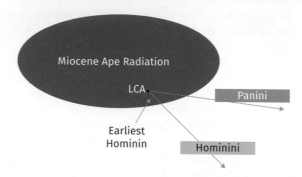

ancestor, we need to find fossils that have a least *one* uniquely hominin character that is shared only with modern humans (see Figure 5.15).

To find this key ancestor we need to keep in mind the characters that define the Hominini: changes in the skull, hip, knee, and foot that indicate a move to bipedalism as the main form of locomotion, an increase in brain size, or changes in the dentition showing a reduction of those features linked to large canines, which would indicate reduced sexual dimorphism. As fossils are found further and further back in time, we look to examine the parts of the skeleton which will give us information about any changes in these three areas.

From hominins in Africa to the first member of the genus *Homo*

For ease, we have organized the discussion of our fossil hominin family into three groups, according to when certain important characteristics first appear. The earliest possible relatives are found **6–4 million years ago (mya)**. These species do not show definitive derived—unique to the hominin lineage—characters, but they reveal early stages of these characters. It is also important to remember that, rightly or wrongly, finding a fossil that is on the human line prompts more academic interest and funding than a fossil that looks as if it is part of the earlier general ape radiation. As a result, scientists are keen to point out anything that is suggestive of a derived character, even when the evidence is not clear!

The second group emerges **4–2 mya**. This group shows important derived characters related to bipedalism, which is why evidence of bipedalism is a defining feature of the hominini. We also see changes in the dentition which many believe are indicative of a reduction in sexual dimorphism—reduced canines. Finally, in Chapter 6, we will look at the period after 2 mya. It is after 2 mya that we start to see major increases in brain size and all the developments that accompany this change, such as tool manufacture and use, meat-eating and group hunting, and a more sophisticated social structure. Many of the earliest fossil hominini have been found in Africa—Figure 5.16 is a useful reference tool in this chapter and the next.

Figure 5.16 The map shows the major sites for earliest hominins, from 7 to 2 mya, discussed in the text.

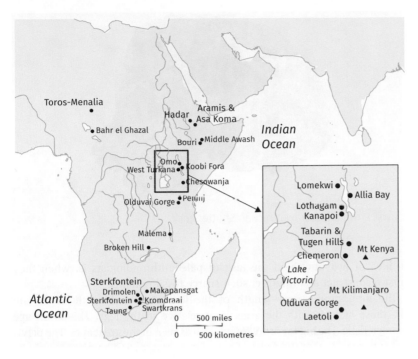

6–4 million years ago: human or ape?

The earliest possible hominin fossils recovered so far have been named *Sahelanthropus tchadensis*. This species is 7–6 million years old and fossils primarily consist of a skull and dentition (see Figure 5.17). It is unusual in that it has a short face, large brow ridges, and small teeth. This makes it different from other Miocene apes, but not like later hominins either. Fossils showing the skeleton below the face (postcranial fossils) of this species are limited but the base of the skull does not suggest the forward position of the foramen magnum that we would expect in a biped (see Figure 5.8).

Orrin tugunensis, found in Kenya and dated between 6 and 5.8 mya, consists of a few teeth and some limb bones. It has large, pointed canines, like those seen in apes. The slightly larger head of the femur has been suggested to indicate bipedality, but this character is also similar to other fossil apes, and in fact to modern chimpanzees. The structure of the neck of the femur has also been claimed to be adaptive for bipedalism, but the same structure is also present in ground-living monkeys.

At between 5.8 and 5.2 mya, in Ethiopia, we find the first occurrence of the genus *Ardipithecus*. *A. kadabba*, as it is named, does not give us any definite evidence about its locomotion, so we cannot say it was bipedal and its teeth are much like those of fossil apes. The second species of this genus, *A. ramidus*, lived a bit later, at 4.4 to 4.2 mya, and shows some of the same dental characteristics—but it does give us some evidence on its locomotion. Only fragments are left of these species—Figure 5.18 gives you

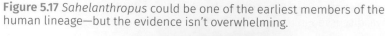

Figure 5.17 *Sahelanthropus* could be one of the earliest members of the human lineage—but the evidence isn't overwhelming.

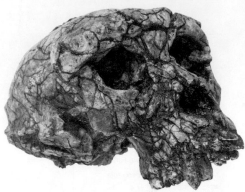

Source: Puwadol Jaturawutthichai/Shutterstock

an idea of just how hard the task of paleoanthropologists is, when they must reconstruct and interpret such scarce fossil remains.

The limb proportions—length of the legs compared to the arms—are like those of apes with the arms long relative to the legs. This is an ape characteristic. The big toe is divergent, also as seen in the great apes. The pelvis seems to possess characteristics of both the great apes and hominins, with the upper part of the pelvis looking as if it has moved somewhat in the direction of bipedalism. It is elongated, like a chimp pelvis, but flared, like a human pelvis. Overall, therefore, *Ardipithecus* shows characteristics that indicate a move toward bipedality, while still showing characteristics that indicate it is moving easily in trees as well. This mix of characteristics suggests that *Ardipithecus* is a good candidate for the earliest recognizable hominin.

4–2 mya: the move to bipedalism

Soon after 4 mya we find fossil ape-like species showing strong evidence of bipedalism, although not the sort of modern bipedal anatomy that we possess today. *Australopithecus* is a genus that we have known about for a long time. In 1925, a fossil child was found by workers in Taung quarry in South Africa. Quarry sites are very hard to date, but it is believed to be about 2.5 million years old. This fossil child is immature—dental similarities with ape growth patterns suggest it was about four years old. Raymond Dart argued that the Taung child provided evidence that the skull would have been positioned on the top of the spine, suggesting it was bipedal, and he described it as belonging to a new hominin species, *Australopithecus africanus*. Later fossils from other South African sites have vindicated Dart's position, as they show a mix of features that include bipedalism along with dental changes of reduced canine size and smaller teeth overall.

Fossils from caves at Sterkfontein and Makapansgat, South Africa produced more australopithecine species (see Figure 5.16). Although most of the fossils recovered are jaws and teeth, there are hip and leg bones as well.

Figure 5.18 When the pelvis of *Ardipithecus* is compared to a modern chimpanzee and human it appears to share similarities with both. The foot has a big toe that seems to diverge even more than a chimpanzee foot, indicating that it is still being used for grasping. The dentition is similar to that of fossil apes and the brain size falls in the modern ape range.

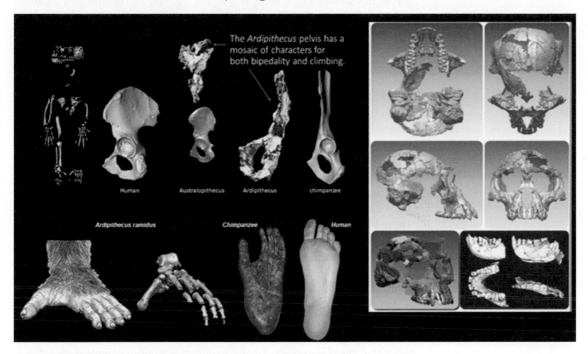

Source: White, T.D., Lovejoy, C.O., Asfaw, B., Carlson, J.P., and Suwa, G. (2015) Neither chimpanzee nor human, *Ardipithecus* reveals the surprising ancestry of both. *Proceedings of the National Academy of Sciences* 112 (16): 4877-4884; DOI: 10.1073/pnas.1403659111

The dentition shows small canines and greatly enlarged molars, features we also see in some fossil apes from the Miocene. Postcranial fossils show clearly that they walked upright, but their relatively short legs, and characteristics in their hands and feet, suggest they still spent much time in the trees.

The limestone deposits in the caves contained the fossils of many animals living at the same time as the australopithecines. Initially scientists thought australopithecines were 'killer apes', hunting and eating many of the animals found in the deposits. We now know, from teeth marks on some of the fossil bones, that they were more likely to be among the 'hunted' rather than the 'hunters'. They were the prey of larger carnivores found at the sites. Many new forms of *Australopithecus* have been recovered in the years since the finding of the Taung child, and with earlier dates. At between 4.2 and 4.1 mya *Australopithecus anamensis* has been recovered from Kenya. The dentition is somewhat similar to fossil apes, but it does provide us with the earliest and best evidence for bipedal walking.

Australopithecus afarensis (3.3 to 2.8 mya) was discovered in the 1970s in Ethiopia (see Figure 5.20). Over three field seasons, the team led by Don Johanson recovered successively a knee joint, then an almost complete skeleton (famously nicknamed 'Lucy'), and finally a collection of individuals. These finds have provided us with a lot of information about the morphology and variation we see within this species. The fossils were compared to earlier dental remains found at a site in Tanzania and they have been grouped together, giving this species a time range of 4 to 2.8 mya. We can determine a lot about the life of *A. afarensis* because the skeleton of Lucy is quite complete. We can see that the pelvis is definitely bowl-shaped, as in modern humans, but the limb proportions and slightly diverging big toe indicate that they did not have a modern gait. The other toes are somewhat curved as well, indicating that *A. afarensis* was still spending a fair amount of time in the trees.

The jaws of *A. afarensis* are robust and the teeth enlarged as in Miocene apes. The canines are reduced in size and it has been suggested that the canine reduction points to a different social structure than that seen in most apes, with reduced male–male competition.

At Laetoli in Tanzania, a set of footprints made by *A. afarensis* show that the foot had both a longitudinal and transverse arch in combination with a separated big toe. This site also preserves footprints of many animals in the 4 to 3.5 million years ago interval. Footprints are a form of trace fossil. These were preserved because a volcano released volcanic ash which covered the area. Animals, including *A.afarensis*, then walked about on the surface and created footprints (see Figure 5.19). These prints were preserved in successive volcanic explosions laying down additional ash. The amazing footprint layer even contains preserved impressions of raindrops!

Another fossil hominin from about 3 mya comes from Sterkfontein (see Figure 5.20). The fossil was encased in limestone deposits and it was through the patient work of palaeoanthropologist Ron Clarke over many years that the skeleton was finally exposed. It is more complete than 'Lucy' and has been given the name *Australopithecus prometheus*. It is very similar to *A. afarensis*.

Figure 5.19 These fossil footprints give us an amazing connection with the early hominins who left them behind—two individuals, one larger and one smaller, walking closely side by side, suggesting a relationship between them.

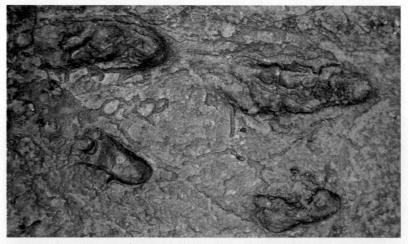

Source: James St. John/Flickr

Figure 5.20 Fossil skeletons of (a) *A. afarensis* ('Lucy') and (b) the very complete skeleton of *A. prometheus* ('Little foot'). Notice that the limb proportions of Little Foot are more human-like, with longer lower limbs, than the skeleton of Lucy.

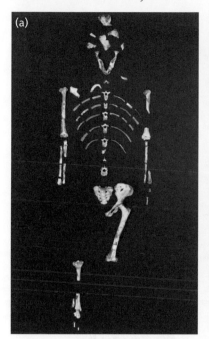 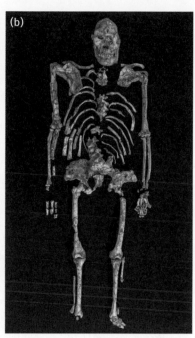

Source: a) Image courtesy of Institute of Human Origins, Arizona State University b) Image copyright Paul John Myburgh (with permission)

A. afarensis and *A. africanus* are grouped as gracile australopithecines. They do not show the heavy skull structure and very large molars seen in the robust australopithecines: *A. robustus* in South Africa and *A. boisei* in East Africa, and an even earlier species at 2.5 mya originally called *A. aethiopicus*. These three species are now grouped in the genus *Paranthropus*. They show increasingly large molars, smaller front teeth, massive jaws, and heavy buttressing of the skull.

From the neck down, both paranthropines and australopithecines have many anatomical similarities. Their brain size is similar as well, and both are in the range of modern apes, but their chewing apparatus sets them clearly apart. Paranthropines have heavy skull bones, with thick sagittal crests on the top of their skulls for the attachment of large chewing muscles. The skull is still small, but the temporalis muscle is so large that extra bony protrusions are developed on the top of the skull so that the muscles have a place of attachment. The jaws are correspondingly large to support both the temporalis and the masseter chewing muscles (see Figure 5.21). All of these heavy chewing architectural features suggest the paranthropines had an abrasive diet of heavy-duty plant material: leaves, slim branches, grass, seeds, and roots. In comparison, australopithecines were eating softer fruits and vegetables, tubers, and possibly incoporating small quantities of meat, such as lizards.

Figure 5.21 The comparison between a generalized gracile australopithecine and a generalized robust paranthropine. Notice the large temporal fossa, sagittal crest, and very large molars and premolars of the robust form.

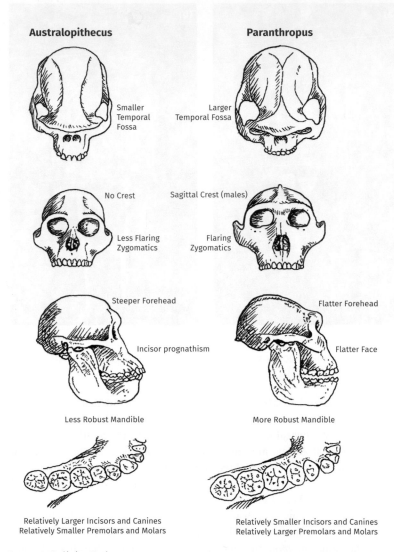

Source: Felicity Baker

Chapter Summary

- It is important to understand that the options available to evolving hominins are a function of our evolutionary past. This is why it is important, in studying human evolution, to be knowledgeable about primate evolution as a whole.

- Primates have retained generalized mammalian features. These are retention of five fingers and toes and of a generalized dentition. Primates are specialized in the mammalian world with a de-emphasis on the sense of smell and increased emphasis on the sense of vision. Importantly, primates replace claws with nails, a factor in our ability to manipulate objects by hand.

- The Last Common Ancestor, LCA, between chimpanzees and humans can be identified by recognizing characteristics that chimpanzees and humans share and then looking for these in the hypothetical LCA.

- Characteristics that are derived in the Hominini are important in identifying the earliest human relatives. These are characteristics related to bipedalism and are found in changes to the base of the cranium, hips, knees, and feet.

- Changes in the dentition of humans include reduction in canine size, changes in palate shape to a parabolic arch, and reduction in the size of molars. Changes in the chewing apparatus in humans involves a reduction in the size of the muscles of mastication, hence a reduction in bony features of the skull.

- Brain size is the feature that allows humans to adapt culturally rather than biologically.

- We identify the earliest members of our human family, the Hominini, by characteristics linked to bipedalism. These changes are found in bones of the base of the skull, hips, knees, and feet. Changes in the hips suggesting early bipedalism are seen about 4.2 mya in *Ardipithecus*.

- Hominins that are clearly bipedal occur in the fossil record around 4 mya in the genus *Australopithecus*. They are small in body size, have large teeth, and their cranial capacity remains small, about the size of modern chimpanzees. At least one branch of this genus gives rise to the hyper-robust *Paranthropus*, who have massive jaws and teeth.

Further Reading

https://www.alltheworldsprimates.org/Home.aspx

Maintained by Primate Conservation, Inc., this is a comprehensive online resource for information about all the living Primate species and subspecies. The database has information on habitat, diet, social behaviour, and conservation.

https://milnepublishing.geneseo.edu/the-history-of-our-tribe-hominini/chapter/primate-classification/

Primate classification and the characteristics specific to each group and species.

Napier, J.R, and Napier, P.H. (1986). *The Natural History of the Primates*, 2nd edition. Cambridge, MA: MIT Press.

This natural history, clearly written by two distinguished primatologists, provides a basic and fully illustrated introduction to the order of primates.

Fleagle, J. (2013). *Primate Adaptation and Evolution*, 3rd edition. Cambridge, MA: Academic Press.

The text of choice for courses on primate evolution. Contains wonderful illustrations and detailed information about modern and fossil primates.

Discussion Questions

5.1   Brain size is the feature that allows humans to adapt culturally rather than biologically. To what extent do other higher primates adapt behaviours that give them survival advantages? What are the long-term evolutionary implications of that?

5.2   What do we mean when we say an animal is specialized or generalized? Is either specialization or generalization more evolutionarily advantageous?

5.3   How do limb proportions, pelvic shapes, cranial capacities, finger and toe proportions, teeth and jaws supply evidence about an animal's locomotion and diet?

6 THE HUMAN STORY SO FAR: THE FOSSIL RECORD

In Chapter 5 we explored our closest relatives, the Primates, and looked at the way we differ from the rest of the Primate order. Understanding which evolutionary changes took place and why they have occurred, allows us to examine the fossil record and to tease out those fossils that represent species related to us. Fossils are fragile and fragmentary. One year we think we have a clear picture of human evolution and then the next year someone finds a fossil that changes the picture. Once a fossil is found there is a process of identification and comparison that can often take years.

An additional problem is that we don't find fossils everywhere—we do not have a complete picture. Early human fossils are found in parts of Africa where sediments of the right age have been exposed (see Figure 5.16). It is important to remember that fossilization is a rare event; most animals that have ever lived have not been preserved. We are looking at a snapshot and a small one at that. Finding those fossils is also a chance event. **Paleoanthropologists**—anthropologists studying human ancestors—travel to those few places where the Earth's sediments have exposed deposits that include the Hominini. Finding a fossil from our earliest ancestors is very much an accidental event! Even so, we have a lot of palaeontological and archaeological evidence to bring to the story.

The early stages of our evolution took place in Africa, and fossil hunters predicted this would be the case and started looking there. Why do you think they might have suspected that this would be the best place to find our earliest ancestors?

Figure 6.1 Olduvai Gorge in Tanzania

The emergence of large-brained hominins

We have seen that bipedalism, a hallmark of the hominin family, makes an appearance in the fossil record very clearly with the genus *Australopithecus*. At the same time as *A. africanus* and the robust paranthropine species are walking around the African landscape, we start to find evidence for a larger-brained hominin in the same deposits, between 2 and 3 mya.

Stone tools and cut marks on the bones of other mammals had been found for many years in Olduvai Gorge (Figure 6.1) and elsewhere, but it was not until 1960 that Louis Leakey, now credited with being the father of modern palaeoanthropology, found sufficient fossil material that he felt comfortable naming a new species. These fossils were sufficiently different from *Australopithecus* that they were collectively named *Homo habilis*. Could this be the maker of the stone tools that were being found?

The evidence for *Homo habilis* is made up of a number of skull and jaw fragments and hand, foot, and leg bones. The skull bones are thinner than those seen in *Australopithecus*, lacking crests and suggesting a larger braincase. But it was not until later that Richard Leakey, Louis' son, discovered a skull, KNM–ER 1470–Kenya National Museum–East Rudolf 1470 when he was working in Koobi Fora in East Africa. This fossil would have had a much larger brain, about 750cm^3 (see Figures 5.13 and 6.2), and the skull did not have the crests present on the smaller-brained hominin

Figure 6.2 Male and female or two species? At the moment, scientists suggest KNM-ER 1470 (left) is *Homo habilis* and KNM-ER 1813 (right) is *H. rudolfensis*.

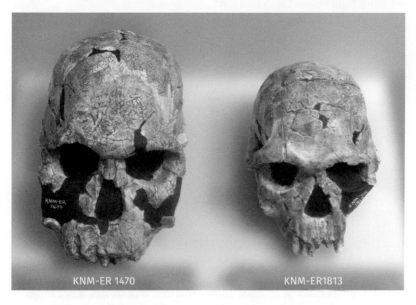

KNM-ER 1470   KNM-ER1813

Source: BY Gunnar Creutz—Own work, CC BY-SA 4.0, https://commons.
wikimedia.org/w/index.php?curid=84448437

species. It has no teeth remaining, but the sockets suggested that they would have been large.

Contrary to the idea that more fossils give us a more complete picture, we often find that more fossils create controversy. Two skulls in particular leave scientists wondering if we are looking at two species or one species that shows size dimorphism based on sex. Look at the skulls in Figure 6.2. Do these skulls represent a male and female of the same species or two separate species? Either could be argued, but at the moment KNM–ER 1813 is put into another species by many scientists. It has been given the name *Homo rudolfensis*.

This flags up a common problem in the study of human evolution, because we are often dealing with very fragmentary evidence. How much variation is too much variation? At what point do we say a fossil is different enough that we must give it a new name? There is no answer as yet to this question, and much of the controversy in palaeoanthropology stems from this issue.

Other species in the *Homo habilis* group include OH-24—Olduvai hominid 24—and many jaws and teeth, some assigned to *H. habilis* (smaller and with smaller jaws and teeth) and some assigned to *H. rudolfensis* (larger, with larger jaws and teeth).

A third early species of the genus *Homo* is also found in East Africa. There are differences between this species and both *H. habilis* and *H. rudolfensis*, but are they significant enough to give it a different name?

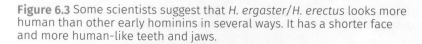

Figure 6.3 Some scientists suggest that *H. ergaster/H. erectus* looks more human than other early hominins in several ways. It has a shorter face and more human-like teeth and jaws.

Source: The Natural History Museum/Alamy Stock Photo

Many scientists think so. Named *Homo ergaster*, it possesses more human-like features of the skull, teeth, and jaws (see Figure 6.3). It is the increase in brain size in particular which marks the emergence of *H. ergaster* about 2 mya. The picture we now have of *H. ergaster* is of a larger brained, fully bipedal hominin, who makes and uses tools. It was the first hominin species to leave Africa—remains have been found in countries as far apart as Israel, Georgia, Java, China, and Algeria.

Case study 6.1
The role of technology

Early hominins began to develop tools which gave them an advantage over other animals at the time in hunting and butchering animals for food. Studying the development of these early technologies helps us deduce our own evolutionary journey.

Stone tools—cutting edge technology

The development and increasing sophistication of stone tools gives us a great deal of information about hominin evolution. Tools are often found in the absence of the fossil people that made them, and give us information on where they lived, how they lived, and how they moved about. Initially, Louis Leakey found tools with the remains of *Homo habilis*, which means 'handy man', a name that implies that *Homo habilis* was the maker of the tools recovered in the same area. These Oldowan tools are found in Africa—the oldest known were found in Ethiopia and are 2.6 million years old!

Initially we thought Oldowan tools were imagined and created by *Homo habilis*, but more recently paleoanthropologists have found similar tools with earlier dates, so we cannot completely rule out *Australopithecus* as the tool-maker. We often find cut marks on animal bones at the same time that we find stone tools. Cut marks demonstrate tool use, even in the absence of tools. The date for the earliest stone tools is regularly being pushed back as new sites are found, but the further back in time we move, the more difficult it is to tell if a piece of broken stone is the result of deliberate human action or just a broken stone!

The main tool associated with *Homo erectus* fossils is the hand axe, also known as bifaces because they have flake scars on both sides. Hand axes are found all over Europe and Africa in some form. In Asia, *Homo erectus* used large flake tools. The way in which hand axes were used is not completely clear, but evidence from experimental archaeology suggests they were used in butchering animals.

Overall, we can say that technology advanced over time, producing better, sharper tools and making better use of materials that were hard to come by. If you have access to flint, obsidian, or any fine-grained material, you want to maximize the cutting edge produced. This seems to be what is happening with the evolution of stone tools. The amount of cutting edge produced per kilo of stone increases over time. There are differences in tool technologies found in Africa from those found in Asia and Europe, and each tool industry contains different tools, although there is some overlap in styles. The Mousterian of Europe, for example, has a specific technique for making tools called Levallois, which was also used by anatomically modern humans during the Middle Stone Age in the Levant. In North Africa, the Levallois technique was used in the Middle Stone Age as well (see Figure A).

Taming fire

The ability to create and use fire was a massive technological step. Fire allowed hominins to keep warm, and to frighten away potential predators. Perhaps most important of all, it allowed them to begin to cook food. Cooked food is much easier to digest, making the energy and materials needed for growth, body maintenance and repair much more readily available. However, evidence for the use of fire is more difficult to be confident about.

Figure A Moving from the Lower Palaeolithic to the Upper Palaeolithic, techniques in stone tool manufacture, as well as materials used, change over time.

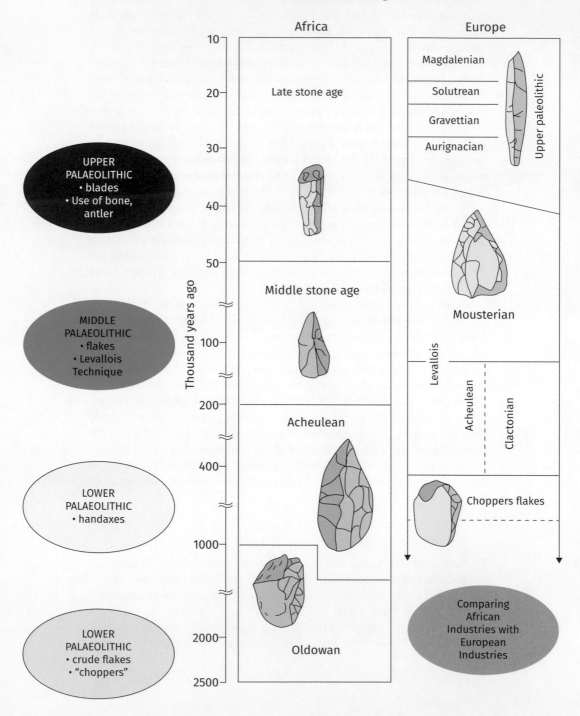

There appears to be evidence of fire at Gesher Benot Ya'aqov in Israel from 790,000 years ago. Evidence from fires includes burned bone, burned stone, and charcoal Unfortunately the site in Israel was open-air, and the difficulty with identifying the controlled use of fire in sites like this is that natural fires can leave 'hearth-like' features. For this reason, fires found at cave sites are given more weight. Such fires would be very unlikely to be a natural occurrence.

Exciting evidence has been found at Wonderwerk Cave in South Africa, where a team from the University of Toronto and Hebrew University have documented the use of fire well back within the cave up to one million years ago. Stone tools were found along with animal bones and wood ash.

Experimental archaeology and taphonomy

Much of what we have learned about stone tools, burned bone, cut marks, and other lines of archaeological evidence comes from **experimental archaeology**. This is a branch of science that tests hypotheses about how tools might have been used by the makers. Specialists have experimented by making copies of tool types and then using those tools for specific tasks: scraping flesh off animal skins, shaping wood shafts, digging, cutting meat from bone, and other possible ways in which the tool might have been used. The wear and polish that collects on the edges of tools can then be examined using scanning electron microscopy (SEM) to quantify what kinds of wear occur for which processes. By comparing the actual artefact with the results of such tests, we get a better idea of how the artefact might have been used.

The discipline of **taphonomy** examines the processes of death. Many different natural processes produce marks on a bone—for example the teeth marks left by non-hominin predators when they kill and eat their prey, or trampling marks made when animals walk over bones in soil. It is important to be able to distinguish cut marks made by a tool-wielding hominin from other damage. Cut marks give us information not only about tool manufacture and use, but also about what the hominin might have been eating.

Cut marks from stone tools are made accidentally during the butchering process, so they are often found where muscle and tendons attach to the bone. Cut marks from tools are also 'V' shaped in cross section when compared to the 'U' shape made by carnivore teeth, due to the shape of the blade. In some cases, cuts on bone can even tell us if the tool user was right- or left-handed. Features called Hertzian cones indicate the direction of cut marks and the 'handedness' of the individual who made that mark can sometimes be worked out from that. It is amazing to be able to imagine ancestral hominins in their daily lives to that level of detail.

Figure B Cut marks like these produced by experimental archaeologists support fossil evidence that our early ancestors used tools to butcher carcasses.

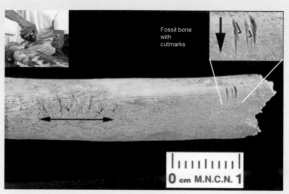

Source: Y. Fernández-Jalvo.

❓ Pause for thought

We find cut marks on bone when we start to find stone tools, about 2.5 mya, but there are suggestions of marks from earlier sites, such as Dikika in Ethiopia, dated to 3.2 mya. If this date is valid, the cut marks would demonstrate the use of tools by *Australopithecus* and some scientists have suggested that fossil remains found at the site are of *Australopithecus afarensis*. Marks that mimic deliberate cuts can be made accidentally, for example by large animals trampling the bones. What other evidence might you want to see to consider if these were true cut marks or cut-mark mimics?

Out of Africa 1—*Homo erectus*

E.B Dubois, a Dutch army doctor, was convinced that the ancestors of man evolved in S.E. Asia. He travelled to Indonesia specifically to find them—and he did! He discovered a fossilized skullcap and femur of *H. erectus* in river gravels along the Solo River. It was clear that the femur was anatomically that of a modern, bipedal person, but the associated skull cap was low and small, with large brow ridges. Dubois named his fossil *Pithecanthropus erectus* (erect ape man). By the mid-20th century, similar fossils had been found in China and Java as well. These are now all grouped together as *Homo erectus*.

One of the confusing problems for students of anthropology is that the names of fossils change over time. This is because as additional fossils are recovered and given new names, it often becomes apparent to scientists that they are all similar enough to be grouped together as one species.

Figure 6.4 A number of different skulls considered to be from *Homo erectus* have been found. They all share the same suite of characteristics, shown in this diagram.

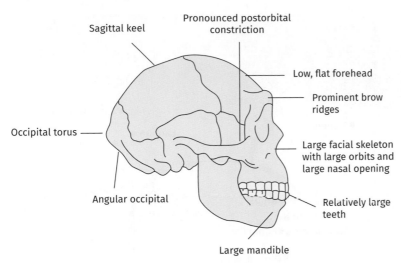

For example, *Pithecanthropus erectus* and *Sinanthropus erectus* are now both considered to belong to *Homo erectus*. The fossils share a suite of characteristics which are now identified and define the group. The brain size is about 1000cm³, and the skull has a low forehead and a long flat cranium. There is a bar-like bony ridge above the eyes—the supraorbital torus—and a constriction behind the torus where the face attaches to the skull. The skull has a thickening of the bone in the centre of the skull called a sagittal keel. It is in the same place as the sagittal crest of paranthropines, but is not for muscle attachment. There is also a thickening bar at the back of the skull in the occipital region called an occipital torus. The jaws project forwardly and there is no chin on the lower jaw (see Figure 6.4) The presence of a chin is an identifying feature of *Homo sapiens sapiens*, modern man, as we will see in more detail later.

Homo erectus fossils of similar age are found in different sites around Africa dated at 1.8–1.1 mya, and at between 2 mya and 1.8 mya, *Homo erectus* is found all over the Old World. While we know there are ancestors of *H.erectus* in Africa, suggesting it evolved there and then migrated out of Africa, we are left with the fact that the dates in Asia are as early as those in Africa ....

The first migration from Africa

Why did *H. erectus* migrate out of Africa? We think *H. erectus* was the first hominin to really engage in hunting, so perhaps they were following migrating herds of game. We find evidence at *H. erectus* sites of more sophisticated stone tool technology, including hand axes, a tool that is not only widespread geographically but spans almost 2 million years. It is the tool that *H. erectus* carried out into the rest of the world (see Figure 6.5).

Figure 6.5 The map shows possible migration routes *Homo erectus* might have taken as it migrated out of Africa and into other parts of the Old World.

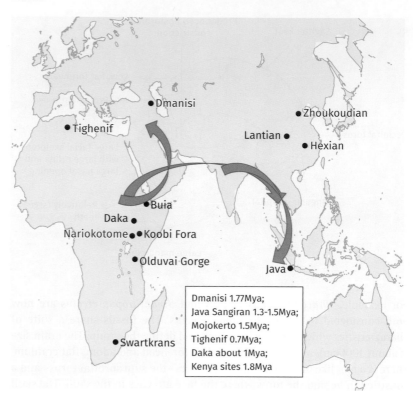

Dmanisi 1.77Mya;
Java Sangiran 1.3-1.5Mya;
Mojokerto 1.5Mya;
Tighenif 0.7Mya;
Daka about 1Mya;
Kenya sites 1.8Mya

It is clear from sites such as Zhoukoudian, a cave site in China, that *H. erectus* also used fire. We know the use of fire goes back at least 500 kya, but as you saw in CS6.1, Wonderwerk cave in South Africa contains evidence suggesting *H. erectus* was able to control fire, if not necessarily to create it, at least one million years ago.

Homo erectus survived in many parts of the world at the same time as other, later developing hominins, including *Homo sapiens*. The fossil evidence shows the species occupied much of the Old World for close to 2 million years. This begs the question about the relationships between the many different species of *Homo* that we will be introducing in the following section. Remember, as with the many forms of *Homo erectus* that eventually became united under one taxon, scientists often disagree about the names and relationships of particular fossils. The variation seen in the skulls found at one site, Dmanisi, in Georgia, tells us a cautionary tale. These fossils are all dated at 1.8 million years old. They all share the suite of *H. erectus* characters we have been looking at, but they show much variation as well, as you can see in Figure 6.6.

Figure 6.6 These skulls are from the same site at the same time in pre-history. The individuals look quite different, but are definitely from the same population. How much variation is needed to call a fossil by a new name?

Source: Virtual reconstruction of the five Dmanisi skulls, with Dmanisi land-scape in the background. © Marcia Ponce de León and Christoph Zollikofer, University of Zurich

Homo heidelbergensis: ancestral to Neanderthals *and* to us?

Out of Africa—again

During the last 600,000 years we find a diversification of possible human ancestors, with both large brains and complex and sophisticated behaviours, but we don't know how they relate to each other and to us. The main issue, from an evolutionary standpoint, is where and how long ago did our modern species, *Homo sapiens sapiens,* emerge? There are several evolutionary models for how this might have happened and the models, and the interpretations arising from them, will change with new discoveries.

For anthropologists, the more general we are in our discussion about the hominins in this period, the more likely we are to be correct! As soon as we start to place specific names on our lineage, we run the risk of getting it wrong. This simply highlights the fact that we are looking at small changes, often occurring in a mosaic way. While we might expect dentition and jaws to change together in the same direction at the same time, because they are both related to eating, we cannot expect jaws and cranial capacity to do so.

As you might expect, when we move forward in time, fossil material is better preserved and there is more of it. Roughly 600 kya a large-brained hominin emerged, now known as *Homo heidelbergensis.* Although they

Figure 6.7 *Homo heidelbergensis* fossils from different places share a similar set of characteristics which are clear in this specimen—notice how very thick the cranial vault bones are, and how the face is robust and projecting.

Source: meunierd/Shutterstock

are found in many parts of the Old World, including Africa, Europe, and Asia, the name originates with a lower jaw from Germany, near Heidelberg. It is this species which may have been the common ancestor of both Neanderthals and modern humans—at 600 kya it coincides with the genetically predicted date of divergence of these two hominins.

What does *H. heidelbergensis* look like? Our expectations are that skulls would become more lightly built, like our own—but this is not the direction human evolution takes here. The Heidelberg jaw, also called the Mauer jaw, lacks a chin and is exceptionally thick and broad. The teeth are somewhat smaller than seen in *H. erectus*. Although they are geographically widespread, they share many similarities. The skulls have long, low braincases with very thick cranial vault bones. But they also show a more rounded shape to the skull in rear view and the sides of the skull—parietal bones—are inflated, with a broad forehead. This gives the skull, in rear view, a pentagonal shape that is human-like. The face is prognathic—with forwardly projecting jaws—the brow ridges are massive and the braincase is larger than that of *Homo erectus* (see Figure 6.7).

Neanderthals and the earliest modern humans

Some of the human fossils recovered in the 19th century were Neanderthals, named after a cave site in the Neander Valley in France where they were found. A partial skeleton was discovered in a quarry in 1856 and was named *Homo neanderthalensis* in 1864 by the Irish anatomist, William King. Earlier discoveries in Belgium and Gibraltar were then included in this taxon. It is currently thought that Neanderthals were evolving in Europe, probably from *H. heidelbergensis* stock, and were well established by 440–400 kya, when Neanderthal fossils were found both in northern Spain and in Kent. Until recently we thought Neanderthals lived around the Mediterranean, but we now know they were more widely distributed. They were adapted to cold climates, such as in England and the Russian Steppes, but have also been found in the warmer, wooded environments that existed in Spain and Italy.

H. neanderthalensis is believed to have lived in groups of about 10 to 30 individuals—for example, a site at El Sidrón Cave in Spain contained fossils of 7 adults, 3 adolescents, 2 juveniles, and an infant. Blood group markers indicate that Neanderthals lived in patrilocal societies: females moved into the group of the males they were mating with. They engaged in group hunting and were highly carnivorous. It is a point of controversy as to whether they made art. Art is generally perceived as the domain of *Homo sapiens,* but *H. neanderthalensis* seem to have made jewellery from animal teeth and eagle talons, and decorated their bodies with red ochre. They inhabited caves and sometimes buried their dead. This is one reason we have so many Neanderthal fossils—but it suggests something even more important. Burying the dead indicates thoughts about afterlife. Some of these burials contain not only the remains of the individual, but remains of animals and tools (see Figure 6.8).

What did Neanderthals look like? It has often been said that if you were standing next to a Neanderthal in an elevator you might think he was a little different looking, but you would not think he wasn't human

Figure 6.8 Notice the circled items on the Kebara burial. On the left is a pig vertebra (above) and some stone tools (below); on the right is a pig molar. Modifications on these 130,000-year-old white-tailed eagle talons from the Krapina Neanderthal site in Croatia suggest they may have been part of jewellery such as a necklace or bracelet.

Source: (a) Credit JAVIER TRUEBA/MSF/SCIENCE PHOTO LIBRARY
(b) © Luka Mjeda, Zagreb

(see Figure 6.9)! Neanderthal bodies were more heavily built than those of modern people, with muscle markings indicating that they were built like a rugby player, although they were shorter than most modern humans at about 1.5 to 1.75 metres tall. They had wide hips and shoulders and weighed between 64 and 82 kg. The skull is differently shaped, as is the brain inside the braincase. The cranial capacity on average is a bit larger than that of modern humans, but that may be the artefact of a small sample size of predominantly males—males are often bigger than females, and this can skew the data in a small sample.

Figure 6.9 A reconstruction of an adult male Neanderthal, from the Natural History Museum in London, UK.

Source: IR Stone/Shutterstock

The emergence of *Homo sapiens*

The first discovery of ancient modern man came from a cave site in France in 1866, shortly after the recognition of Neanderthals in 1864. Three human skulls were found with advanced stone tools from the Upper Palaeolithic, along with perforated shells to use in jewellery. The site, Cro-Magnon (which means 'Big Hole' in French), gave its name to the fossil inhabitants. Although close both geographically and in time to the latest Neanderthals (at La Ferrasie), they are quite different anatomically and look very much like modern people. The name 'Cro-Magnon' is used to identify anatomically modern people whenever we see them in the fossil record. You can see from Figure 6.10 that they are taller and more slender, believed to be an adaptation to a warmer climate. This suggests *H. sapiens* evolved in a warmer climate and migrated into colder areas inhabited by *H. neanderthalensis*. They would have needed clothing of some sort, and to be able to use fire effectively. There are also significant differences in the skull, also seen in Figure 6.10. The cranial capacity is roughly the same (possibly somewhat smaller), the shape of the skull is different, the face is flatter and shorter, and there is a chin on the mandible.

Behaviourally they seem different too. *H. sapiens* appears to have lived in larger groups than *H. neanderthalensis*, although still hunting for much of their food. Art starts to appear, with the earliest known cave art at about 43 kya in Spain and 40 kya in Indonesia. Use of red ochre, possibly for bodily adornment, is known from South Africa about 70 kya. Other forms of engravings also appear on bone, antler, and ivory. Figurines and three-dimensional art appear. *H. sapiens* seems to have made widespread use of shells and teeth for jewellery.

We have good evidence for modern humans living in the same areas as Neanderthals, and DNA evidence supports interbreeding between the two groups. We have evidence for people similar to the Cro-Magnons in this period from North Africa, China, and the Levant—which includes present-day Syria, Lebanon, Jordan, Israel, and Palestine. The question arises, how far back in time can we extend the line of modern-looking people? Where do we find the earliest evidence for anatomically modern humans? There are several theories about where modern people first arose and how. As you might imagine, it often depends on the identifying features seen in Figure 6.10, and the interpretation of these characteristics can be controversial. *H. heidelbergensis*, the fossils discussed in the previous section, are believed to have given rise to both Neanderthals and to these later people, who we find at about 300 kya in East Africa, 450kya in Morocco, and later in South Africa and the Sudan. This is the group of fossils widely considered as the earliest known fossil *Homo sapiens*, any of which could be ancestral to modern people. These fossils are also associated with Middle Stone Age tools that are very sophisticated.

The earliest site with anatomically modern humans is Jebel Irhoud in Morocco, which was very recently re-dated to a much earlier date of between 350,000 and 280,000 years ago. If correct, this new fossil evidence pushes the earliest examples we have of the *Homo sapiens* lineage back by more than 100,000 years.

Figure 6.10 Neanderthals were shorter and stockier than modern people, adapted to a colder, harsher life. The limb bones are thick, with heavy muscle markings. The skulls of *H. neanderthalensis* and *H. sapiens* differed too. The yellow shapes indicate the shape of the skull in rear view.

(a)

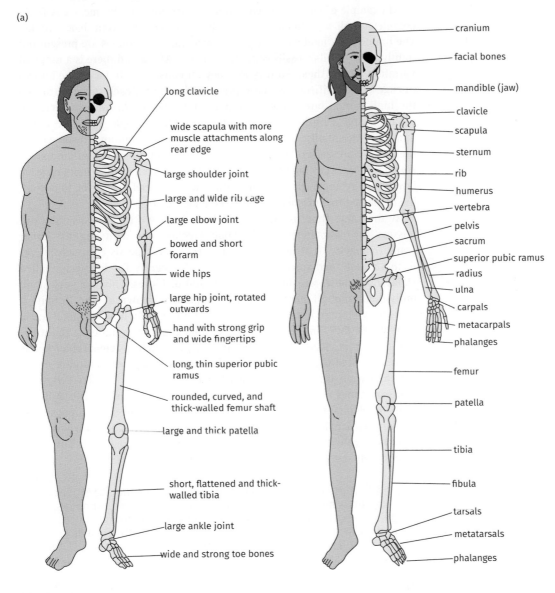

(b)

Neanderthals:
Long, low skull shape
Low forehead
Double-arched browridges
Projection of the nasal bones
Sloping and projecting maxilla and zygomatics
A space after the third molar
Oval skull profile in rear view
Occipital 'bun'
1450 cc cranial capacity no chin

Homo sapiens:
High, globular skull shape
High forehead
Very small browridges
Smaller nasal bones
Pushed inward maxilla and zygomatics, creating a sharp
Hexagonal skull profile in rear view
Rounded 'bun'
1400 cc cranial capacity chin

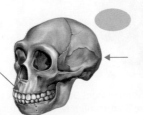
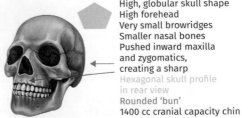

It is important to note that although the date of the earliest *sapiens* is 350–280 kya and the date for Neanderthals is 440 kya, because they share a common ancestor the earliest humans must be at least 440 kya. This is a good example of what a shared common ancestor actually means. A fairly complete skull was found along with other remains from Jebel Irhoud. The braincase is modern in shape and size and brow ridges are present but lightly built. Similar fossils come from across Africa and there is a range of variation among them—so why are they all considered to be on the human lineage and classified as *Homo sapiens*? Individually, each specimen shows the high frontal bone, large globular skull, reduced brow ridges, cranial capacity, and dentition that continue in modern people. The face is flat and short, essentially 'tucked in' under the rounded cranium. In some cases there is the presence of a chin, which is not present in any hominin except *Homo sapiens*. The chin seems to be a structural remnant of a more massive jaw, which remains when much of the jaw is reduced. It may be important in supporting the incisors in a bite (see Figure 6.10).

At the same time, in South and East Asia, we find fossils that still possess the characteristics of *Homo erectus*, but with slightly larger cranial capacity. It would appear that *H. erectus* persisted much longer in this region, perhaps as late as 70 kya. A recently found species called *Homo floresiensis* was found on the island of Flores. This is a very small individual, about 1 metre in height, popularly known as 'the Hobbit'. It looks very modern in many ways: slender build, fully bipedal, small teeth, human-like face, and tools. But the cranial capacity is 420cm³, about the size of a chimpanzee. It is dated at 17 kya, which is remarkable! It has been suggested that this species evolved from *Homo erectus* living in the area and went through a process of size reduction, or species dwarfism, that sometimes results when animals are limited to an island life. A dwarf form of elephant, *Stegadon*, was recovered from the same deposits as *H. floresiensis*, along with stone tools.

In the Levant, the picture is even more complex. This region was occupied successively by both Neanderthals and by *Homo sapiens*. This may be

Figure 6.11 These fossils represent two of the earliest members of *Homo sapiens*, but as neither of these skulls has a lower jaw, we cannot assess the presence of a chin. The earliest (a) is Jebel Irhoud from Morocco at 350 kya and the second (b) is the Florisbad skull from South Africa at 259 kya. Although there is variation and they both seem to be quite robust, they also share the suite of characters that identify them as *Homo sapiens*.

Source: a and b: Ryan Somma from Occoquan, USA, taken at the David H. Koch Hall of Human Origins at the Smithsonian Natural History Museum

because the Levant is geographically positioned as a corridor through which Neanderthals might have moved south from Europe and early *H. sapiens* might have migrated north from Africa. We know that people moved about a lot at this time, following migrating herds of game or moving when the weather became too severe to find sufficient food.

The latest known geographic and temporal range of *Homo erectus* and *Homo neanderthalensis* coincides with the appearance of *Homo sapiens* and the migration of *H. sapiens* into both Europe and Asia. Neanderthals survived until 30 kya in Europe and *Homo erectus* in Asia until about 100 kya. This means that there is a long period of overlap between all these fossil species, both in where they were living and in the time they were living there (see Figure 6.12).

What does it all mean?

The study of our human origins, and the theories about our evolution have changed dramatically over the last 100 years. What drives this change? It is partly the result of new techniques such as DNA, X-ray, and computer statistical analyses, along with new dating techniques. This makes it possible to review fossils discovered during previous generations and put them into a new context. But the biggest factor in our ever-changing view of our past comes from the discovery of new fossils. When a new fossil is discovered the entire set of hominin species needs to be re-examined, and the fossil either placed into accepted scenarios or new concepts developed to accommodate the fresh information it provides. This can take many years, and agreement among scientists even longer! We have presented a general view

Figure 6.12 The map shows the migration route of *Homo sapiens* out of Africa, and the dates when their earliest fossils are seen in other parts of the world. It also shows the range of *Homo erectus* and *Homo neanderthalensis* during this time.

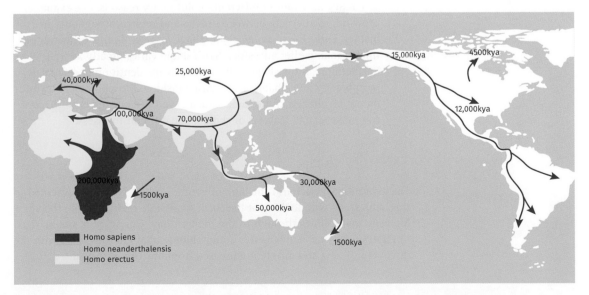

of the players in our human evolutionary story, but you can expect the evolutionary scenario to continue to change, possibly in surprising ways.

The story we can tell from the fossil record at present is one with many unknowns. But the methods we use to examine the fossil record will continue to be the same. The main ingredients in determining relationships among fossil species include the consideration of several factors, which scientists try to determine for all fossils. These are:

- *Time*: What is the date of the fossil? While this sounds easy, it is not always possible to find good material to date. Different methods are used at different sites depending on the sediments in which the fossil is found. This means dates are usually taken with some caution and they frequently change as new and improved methods of dating are invented.

- *Space*: Where was the fossil found? What was the geologic setting at the time the species existed? Where does the site sit geographically and what is the relationship with other sites? We also need to consider that important fossils may have not been found, simply because the sites are not exposed. For example, some people believe that much of human evolution has taken place in coastal areas. We do not find fossils in these areas because the potential sites are under water. Consequently, we can speculate about the importance of coastal habitats to human evolution, but we have no evidence for it.

- *Morphology*: What does the fossil look like, and what does its functional morphology tell us about how it lived in its environment? This is what the bulk of this chapter has been about. You may have received the impression that assessing a fossil's morphology and comparing it to other fossils is a straightforward process, but for many of the fossil hominins there is not always scientific agreement on what the morphology is and what it means. We expect that our evolutionary ancestors and cousins have changed in different ways at different times and for different evolutionary reasons. For example, brow ridges alternate between gracile and robust throughout our history. Early *Homo sapiens* often have large and thick brow ridges, thicker than earlier ancestors. Skull shape varies greatly as well. Furthermore, brow shape and features of the dentition or cranial morphology may change in a mosaic fashion. Morphological features do not all change at the same time and in the same direction. This can make simple features of skull shape difficult to assess.

The story so far

The evolutionary scenario for human evolution that we have presented in this book starts with looking at the earliest possible relatives from the latest Miocene and follows through to the modern day. Our story is summarized in Figure 6.13—but remember, it could change again at any time, with the discovery of new and different ancestors from our complex past.

Figure 6.13 This timeline shows the species discussed in the text, their main defining characteristics, and the time range they occupy. Notice that many of the species are marked in pink and defined as having insufficient evidence.

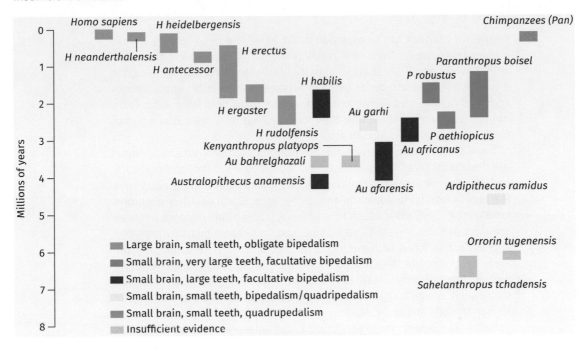

The bigger picture 6.1
The evolution of language

Speech and complex language are regarded as one of the defining features of modern humans—*Homo sapiens sapiens*. It may therefore surprise you to learn that there is no adaptation in our bodies dedicated solely for language. Structures similar to our brains, mouth, teeth, tongue, face musculature, larynx, and hyoid bone are found in many animals that do not produce speech. All have been evolutionarily modified to promote our ability to communicate in an infinite number of ways about an infinite number of subjects—we even talk about things that do not exist. This is something that—as far as we know—no other creature can do, and questions about how and why we evolved this amazing attribute are as yet unanswered. We do however have many lines of inquiry, and it is these sources of information that we discuss here.

What is language?

Before we can look at the evolution of language, we need to define what human language is. There are certainly forms of communication in the animal world which are used to impart complicated information. Bees, for example, use flight and dance patterns to let other bees know where a novel source of food is—and even that it might not be ready until later. But they do not discuss theories about how their food evolved or suggest new types of food. This is a uniquely human activity.

There are also many non-verbal forms of communication and the ability to use these is inherited. Many animals have 'words' by which they communicate, and in a few they may be compound words; chimpanzees and gorillas have learned several hundred words and expressions. Gestures, body positions, facial expressions, eye-contact, tone of voice, posture, and distance between speakers are all integrated into communication systems by many primates. There are also forms of communication that don't require sound, but the ideas expressed are the same as if they were spoken. Sign language, drum signals, fire beacons, semaphore, and Morse code are all non-verbal, but many, such as sign language, contain the ability to express the full range of human thought and expression.

There are elements we expect from human language that make it special. Although many animals can be shown to communicate effectively, we humans have taken this to another level. We form intentional and arbitrary links between form and meaning and agree upon these links within our groups. These are **symbolic links** and they are the basis of human language. Is grammar necessary for meaning? Not always ...

No shirt. No shoes. No entry.

But grammar becomes necessary for when language becomes more complex. The ability to put thoughts within thoughts allows infinite meanings, because it allows us to embed phrases within phrases and sentences within sentences as in

Tom met Yasmin, who is his friend's sister.

The origins of language

When we start to study the origins of language, we find several lines of evidence. They include:

Genetic evidence

Is there a speech gene? One has not been found, but there is a speech-related gene called the FOXP2 that is present in different variants in the Hominoidea. Humans have two fixed variants, one shared with orangutans and the other

Figure A FOXP2 variation in apes and humans.

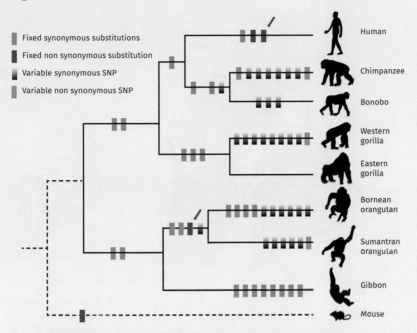

Source: Staes, N., Sherwood, C.C., Wright, K., de Manuel, M., Guevara, E.E., Marques-Bonet, T., Krützen, M., Massiah, M., and Hopkins, W.D. (2017). Nature https://www.nature.com/articles/s41598-017-16844-x#Abs1

with other very distantly removed mammals, such as mice. This would produce an unusual **cladogram**—from the Greek for clados 'branch' and gramma 'character', a diagram used in cladistics to show relations among organisms. It would suggest that with respect of this character, we are more closely related to orangutans! Mutations in this highly conserved FOXP2 gene might have enabled the very fine oro-facial movements needed for speech and could have triggered the development of language through epigenetic change—see Figure A.

Morphological evidence

Morphological evidence is derived from modern humans and apes. When we look at the anatomy of the larynx in apes and humans, we find that in a newborn the situation is quite similar to that seen in chimpanzees. The larynx is relatively high in the throat and the epiglottis is a relatively long way away from the palate. Having it close to the trachea prevents children from choking when they nurse, but as they grow older, the larynx drops and the epiglottis assumes the adult position. From about 4 to 6 months old a child can start making the sounds of speech, but they will also be in a position to choke on food.

Figure B Morphological evidence of the anatomical changes associated with human speech.

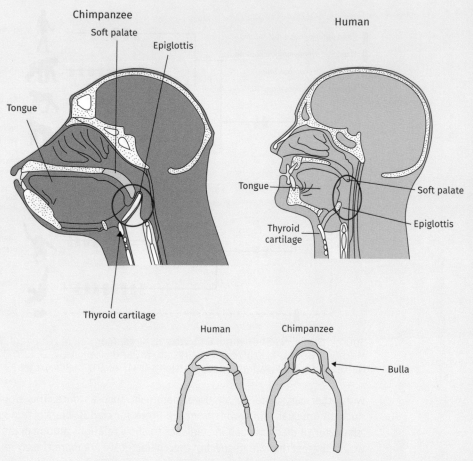

Speech production preparation
Larynx descent and speech

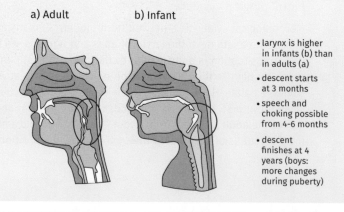

- larynx is higher in infants (b) than in adults (a)

- descent starts at 3 months

- speech and choking possible from 4-6 months

- descent finishes at 4 years (boys: more changes during puberty)

Eventually the speaking ability improves and the choking risk diminishes. The hyoid bone is a free-floating bone in the throat. It moves lower in the throat as the larynx moves lower. The **hypoglossal nerve** is the motor nerve of the tongue, and as you might expect, it is larger in humans than in other apes as the tongue is capable of the many controlled movements needed for speech. This nerve must pass from its origin in the brain, through the **hypoglossal canal** and ultimately to the tongue. Like the nerve, the canal through which it passes is correspondingly larger than that seen in other apes.

We know the brains of *Homo erectus* became much larger than those of early hominins, growing from 1000 cm^3 to 1200 cm^3. There must have been an evolutionary advantage to this, because the brain is a metabolically expensive organ to produce and run. Neanderthals have brains that might even have been larger than ours (1450 cm^3). As you have seen, the brain did not evolve specifically for speech, although studies have shown that there are parts of the brain that have been reorganized to carry out specific language functions. Furthermore, if damaged, different parts of the brain can take over language functions. But the brain has not increased in volume over the last 500,000 years. There is a trade-off between brain size and bipedalism. Humans are walking creatures, so the pelvis has an ideal size and shape to allow us to move efficiently on two legs. If the brain (and the skull that protects it) gets too large, the infant would get stuck in the birth canal. This would kill both the mother and the infant, and is very strongly selective for keeping brain size within a particular range.

Fossil evidence

The fossil evidence for the development of speech is fragmentary at best. We can look at the cranial material, jaws, teeth, and hyoid bones and compare them with the condition seen in modern apes and humans. But as you know, hominin

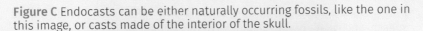

Figure C Endocasts can be either naturally occurring fossils, like the one in this image, or casts made of the interior of the skull.

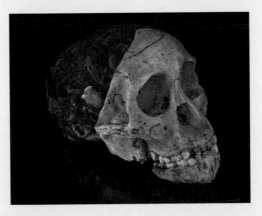

Source: By Didier Descouens—Own work, CC BY-SA 4.0, https://commons.wikimedia.org/w/index.php?curid=29379536

fossils are not common. We rarely get natural **endocasts**—a fossil cast of the inside of the cranium (Figure C)—and in any case, the evidence of brain size increase alone does not indicate language. The size of the brain/skull during human evolution may have been more strongly related to an increase in protein in the diet rather than to changes in intelligence or the development of speech.

Another piece of anatomy that can be brought to bear on the question of speech and language is the hyoid bone. This bone sits in the throat and does not articulate with anything, but it forms a site of muscle attachment for several muscles of the mouth, throat, and skull. Fossil hyoids have been found, although they are rare. When a Neanderthal hyoid is compared with that of a chimpanzee and a human, it is identical to that of humans.

Archaeological evidence

Finally, we need to look at the archaeological evidence, which enables us to quantify the cultural developments suggesting language was a factor. When *Homo erectus* migrated out of Africa, they carried with them a complex set of behaviours in the form of technology, and lived in groups. Since humans use language to pass on cultural information about hunting practices, making tools, and determining group movements, there is a strong suggestion that language was present at that time. Most anthropologists consider that the most important factor in the evolution of speech is the social environment and how it changed during human evolution. The size and complexity of social systems, together with its importance in hunting and culture, were probably the main factors promoting development of language, much more important than morphological change in the larynx, and bipedalism.

❓ Pause for thought

How does the archaeological evidence supply information about human speech, and what are the limitations of this evidence?

Summarizing the model

The discoveries of fossil humans show that brain size increase began about three million years ago with the earliest species of *Homo*, *Homo habilis*, but a greater increase occurred with the emergence of *Homo erectus* just after two million years ago. It was not until just over half a million years ago that the final and most dramatic increase occurred with the appearance of *Homo sapiens*, including Neanderthals. We believe that both brain size increase and the beginnings of symbolic speech occurred at the time of *Homo erectus*. This was the fossil species that first left the woodlands for open country savanna habitats away from the protection of trees, the species which dispersed out of Africa, travelling across many different habitats in Europe and Asia, crossing large rivers and even the open ocean to get to Java, and finally this was the species which learned to control fire, both for protection and for cooking food such as meat. All of these factors suggest that *Homo erectus* must have been able to communicate at a much higher level of sophistication than we see in non-human primates. Later hominins like Neanderthals had large brains and the capacity to survive in cold conditions, and very late *Homo sapiens* had developed high technologies and art. Our abilities to think and speak are still defining characteristics today.

Chapter Summary

- *Homo habilis*, the first member of the family *Homo*, appears a little over 2 million years ago and is followed quickly by *Homo erectus*. *Homo habilis* shows an increase in cranial capacity to a range between 550 to 680 cm³. *Homo erectus* continues that trend, with cranial capacities in the range of 880 cm³ to over 1200 cm³ in the latest members of the species.

- Stone tools start to appear in the archaeological record about 2.5 million years ago and are coincident with both *Homo* and *Australopithecus* species. Stone tool technology develops quickly and is believed to be what allows *Homo erectus* to be the first hominin to migrate out of Africa and into Europe and Asia. It persisted in Asia until very late (70 kya).

- *Homo neanderthalensis* and *Homo sapiens* are thought to have diverged from a *Homo heidelbergensis*-like ancestor about 600 kya, according to evidence of modern genetic diversity. Neanderthals and early humans both had sophisticated toolkits and made jewellery. Neanderthals buried their dead and both used fire.

- *Homo sapiens* don't appear in the fossil record until about 350 kya, but we believe they must have been present earlier. Upper Palaeolithic tools show a great diversity of tool types, made from a variety of materials. Sometime after 100 kya, *Homo sapiens* from Africa left on a second exit from the continent and are believed to have supplanted Neanderthals, and remnant *Homo erectus* populations in the Old World.

Further Reading

https://humanorigins.si.edu

An excellent website covering all the aspects of human evolution mentioned in this chapter, with an emphasis on current research.

nhm.ac.uk/discover/human-evolution.html

The Natural History Museum supplies a 7-million-year walk through time. It includes excellent photographs, good models, and information on ground-breaking research.

https://www.biointeractive.org/classroom-resources/human-origins

An interactive resource with information on palaeobiology, human evolution, and anatomy.

Lewin, R. (2004). *Human Evolution, an Illustrated Introduction*, 5th edition. Oxford: Wily-Blackwell.

The text places human evolution in the context of humans as animals, while also showing the physical context of human evolution, including climate change and the impact of extinctions.

Stringer, C., and Andrews, P. (2011). *The Complete World of Human Evolution*, 2nd edition. London: Thames and Hudson.

A very readable overview of human evolution with accessible text, filled with diagrams and photos.

Andrews, P. (2016). *An Ape's View of Human Evolution*, 1st edition. Cambridge: Cambridge University Press.

A more in-depth examination of the earliest human ancestor that promotes a view from our ancestral apes, with details on fossil apes, palaeo-environment, functional morphology, and paleontological methodology.

Discussion Questions

6.1 As mentioned in the text of this chapter, scientists do not always agree on the morphology of many of the fossil hominins, and what it means—yet they are all studying the same fossil remains. Suggest reasons for these scientific conflicts.

6.2 How do humans compare to other primates in terms of their genetic diversity? How do they compare to other mammalian species? What does the level of diversity seen in humans suggest about our evolutionary past?

6.3 Fossils can only tell us so much. Suggest some of the limitations of the fossil record.

GLOSSARY

Adaptive radiation The rapid evolution of new forms of a common ancestor, often colonizing different environments by evolving adaptations for those environments.

Adduct To move a limb towards the midline of the body or towards another part: the Human big toe is adducted. It is the opposite of abducted.

Allele Variant of a gene with some differences in the DNA base sequence.

Allopatric speciation speciation where two populations have become geographically separated.

Anthropogenic Changes to the environment caused by human activity.

Batesian mimicry A form of *mimicry* in which a harmless species has evolved to imitate the warning signals of a harmful species, which are directed at a predator of them both. It is named after the English naturalist W. H. Bates, after his work on butterflies in the rainforests of Brazil. In Müllerian mimicry, harmful species imitate each other to provide added protection.

Behavioural systems How the behaviour of organisms can affect their interactions with other organisms in their environment, in ways that affect their survival.

Biological fitness The ability of an organism to survive, reproduce, and pass on its genome to future generations.

Biological species organisms which are capable of interbreeding to produce fertile offspring.

Biometrical genetics The inheritance of quantitative characteristics, which are caused by the interactions of many genes with the environment.

Bipedal A form of locomotion using two legs. This is novel among primates, used only by Humans.

Character displacement The phenomenon in which differences among similar species whose distributions overlap geographically are accentuated in regions where the species co-exist, but are minimized or lost where the species' distributions do not overlap.

Chemosynthetic bacteria microorganisms that derive energy from the breakdown of organic or inorganic compounds without involving light.

Chromatin remodelling Changing the arrangement of histone proteins surrounding DNA, to alter the activity of protein-coding genes.

Circadian clock A biochemical process that is synchronized with the patterns of day and night. It controls the activity of other biochemical processes.

Clade see Cladistics.

Cladistics The approach to biological classification in which organisms are categorized in groups, or clades, based on the most recent common ancestor.

Cladogram From the Greek clados, 'branch', and gramma, 'character'. A diagram used in cladistics to show relations among organisms.

Conservative The retention of earlier characteristics or morphologies. Primates are skeletally conservative, retaining the original five mammalian toes and fingers, for example (also known as primitive characters).

Conserved (genes) A gene with a nucleotide sequence that has remained unchanged during evolution. It plays a unique and essential role in the development of an organism.

Cyanobacteria photosynthetic prokaryotes which can fix both carbon and nitrogen.

Cryptic species These occur when morphologically almost identical organisms are reproductively incompatible.

de novo Means 'from the beginning'. Primate characteristics have a history from a common ancestor and it is from this that evolution makes changes.

Dental arcade The shape of the palate and the mandible. In the apes the palate can be squared or parabolic (rainbow shaped).

Dentition Teeth.

Derived A changed morphology or characteristic that moves away from the original type.

Derived characters Characters that differ from a named common ancestor. Shared derived characters (also known as synapomorphies) form the basis for cladistics.

Diapause a dormant stage in the arthropod life cycle which takes place during unfavourable environmental conditions.

Directional selection When natural selection favours an extreme phenotype over other phenotypes. This causes the allele frequency to shift over time in the direction of that phenotype.

Disruptive selection When natural selection selects against the average phenotype in a population. Most individuals in this population will have the extreme phenotypes at either end of the range.

Distal Refers to a part of the anatomy that is positioned far away from the centre of the body or the point of its attachment—the opposite of *proximal.*

DNA fingerprinting (profiling) A process in which electrophoresis is used to separate DNA fragments in order to create a distinct pattern of bands like a bar code. DNA profiles of different species can be compared in order to build a picture of evolutionary relationships.

Double circulation An arrangement of arteries, veins, and the heart. The blood passes twice through the heart on its complete journey round the body.

Endemic native only to a particular geographical area.

Endocasts A cast created inside a fossil skull that gives an estimate of the brain size and morphology. It is formed by mineralization that takes place in the cavity left after the brain has deteriorated.

Endosymbiotic a relationship in which an organism lives inside another one, and gives its name to the theory that mitochondria and chloroplasts originated as symbiotic bacteria.

Epigenetics The addition of chemical markers to DNA or chromosome proteins that change the expression of protein-coding genes. These markers do not alter the nucleotide sequences of genes. These markers may be passed on to offspring, in some organisms.

Epistasis The interaction of different genes. The term is most often used to describe the suppression of the effect of one such gene by another.

Evolvability The ability of organisms to adapt to their environments through natural selection. This involves changes and rearrangements to their genomes.

Experimental archaeology The branch of archaeology that uses experiments to study the archaeological record. By creating modern comparative data sets, the manufacturing processes, materials, and uses of archaeological material can be better understood.

Femur The thigh bone, which articulates with the hip at the proximal end and the knee at the distal end.

Foramen magnum 'Large hole'. In anatomy, this refers to the large opening at the bottom or back of the skull from which the spinal cord exits the braincase.

Founder effect Occurs when a small group of individuals becomes separated from the parent population. The genetic diversity of the new population is smaller than the parent population.

Gametic isolation Occurs when there is little or no interbreeding between organisms of similar species due to incompatibility of the gametes.

Gel electrophoresis A laboratory technique using an electrical current to separate DNA, RNA, or protein molecules based on their size and electrical charge, which affects the rate they move through a gel.

Gemmules Darwin's name for minute particles of inheritance that he hypothesized are shed by all cells in an organism and circulate throughout the body, collecting in the sex organs to be passed on to the offspring.

Gene A sequence of DNA which occupies a particular location on a chromosome and codes for one or more proteins.

Gene pool The set of all genes, or genetic information, in a population of a particular species. The larger the gene pool, the greater the genetic diversity.

Germ cell A cell that divides to produce a gamete for sexual reproduction.

Gondwana An ancient supercontinent of the Southern hemisphere.

Habitat isolation Organisms that live in different habitats so do not have the opportunity to mate and reproduce.

Haplotype A sequence of DNA nucleotide bases inherited as a unit. These may affect the production of proteins (alleles of genes) or they may be in non-coding regions of DNA. They are useful for the identification of individuals.

Homeotic gene These are sequences of DNA that produce proteins that regulate the activity of other genes involved in the development of organisms.

Horizontal gene transfer The movement of genetic material between organisms other than by sexual or asexual reproduction, eg bacteria share genetic material in plasmids leading to the rapid acquisition of antibiotic resistance.

Hypoglossal canal A foramen, or hole, in the occipital bone of the skull, through which the hypoglossal nerve travels from the brain to the tongue. Its size indicates the size of the hypoglossal nerve.

Hypoglossal nerve The motor nerve which innervates most of the muscles of the tongue. Its size is an indication of the complexity of the muscles of the tongue.

Limb proportions The relative lengths of the arms and legs. These proportions vary in primates and indicate how the animal locomotes.

Mandible The bone in the lower jaw.

Mechanical isolation When organisms cannot reproduce as a result of incompatibility of the reproductive anatomy.

Meiosis The type of cell division that produces haploid gametes from diploid cells.

Methanogenic the ability of some microbes to create methane.

Microsatellite DNA A sequence of DNA nucleotide bases that show characteristic sequences (eg CGCGCG), repeated up to 50 times. Microsatellite DNA sequences do not code for proteins and are used in genetic fingerprinting techniques.

Mitochondrial DNA also known as mtDNA or mDNA and is found only in mitochondria.

Mitosis The type of cell division that produces daughter cells that are genetically identical to the parent cell.

Modern Synthesis The revised theory of evolution published in 1942 that reconciles Darwin's theory of evolution with Mendelian genetics.

Molecular clocks A technique that correlates the accumulation of changes (mutations) in proteins of DNA with time, to estimate when two groups of organisms last shared a common ancestor.

Monograph A detailed written account of a single, very specialised topic eg a single species of organism.

Morphological species is a way of defining species based only on appearance.

Morphology The study of shapes. Morphologies refer to the colours, shapes, forms, and types of variations seen in animals and plants.

Mosaic Evolution The idea that evolutionary change takes place in some body parts or systems without simultaneous changes in other parts and at different rates of evolution.

mRNA Messenger RNA carries information from the DNA molecules in the nucleus to the site of protein synthesis at the ribosomes.

Multilevel selection How the phenotype of an organism is affected by different levels of organization (eg the ecosystem, population, cellular, and genetic levels). It looks at interactions between different levels of organization.

Mutation A change in the amino acid sequence of a protein, the number and structure of chromosomes, or the nucleotide sequences of DNA.

Niche construction How an organism alters the environment to enhance its own survival, sometimes at the expense of other species.

Nonoverlapping magisteria An idea proposed by Steven Jay Gould to try to reduce the supposed conflict between science and religion.

Ontogeny The study of the developmental (anatomical, behavioural) stages of an organism from fertilization to maturity to death.

Outbreeding occurs when a population of a species breeds with individuals from another population.

Paleoanthropologists From the words paleo, 'ancient', and anthropology, 'study of man', these scientists study the fossil evidence related to human evolution.

Pangaea a supercontinent that came into existence during the Late Palaeozoic Era.

Pangenesis Darwin's term to describe his hypothesis that all the body parts of the parents could contribute hereditary material to the development of the offspring by shedding gemmules which collect in the sex organs. Any effect of the environment on the parent during its lifetime could be passed on via the gemmules.

Parabolic A term describing the shape of the dental arcade in humans.

Parietal bones Pair of bones on either side of the braincase; they form the cranial vault and meet in the middle at the sagittal suture.

Phenotypic plasticity The way a genome produces different phenotypes in response to changes in the environment.

Phylogentic tree a branching diagram showing the evolutionary relationships between species.

Polygenic A group of genes that interact to affect a phenotypic trait. This creates many possible allelic combinations, so often leads to continuous variation.

Polymorphic A population in which there are at least two variants of a particular DNA sequence. The most common type of polymorphism involves variation at a single base pair. Polymorphisms can also involve multiple genes.

Polyploidy the non-disjunction of chromosomes during cell division, resulting in additional sets of chromosomes.

Postcranial Refers to all the bones of the skeleton below the head, or cranium.

Postzygotic and Prezygotic isolation two types of reproductive isolation. Postzygotic can be caused by the non-viability of the zygote or sterility of the offspring. Prezygotic describes mechanisms which prevent gametes meeting. see Genetic Isolation.

Prognathic From 'pro' for 'forward', and 'gnathic' for 'related to the jaw', this term indicates a forwardly positioned jaw. Alveolar prognathism: forward projection of the dental alveolus; midface prognathism, projection of the face.

Proteome set of proteins produced by a given organism at a given time and under a given type of conditions.

Pulmonary circulation The part of the circulatory system that carries deoxygenated blood away from the right ventricle of the heart to the lungs, and returns oxygenated blood to the left atrium of the heart.

Quadrupedal A form of locomotion in which the animal uses all four legs.

Retrovirus A virus that inserts a DNA copy of its RNA genome into the chromosomes of a host cell.

Sagittal crest A bony ridge following the line of the sagittal suture on the top of the skull which provides a bony surface for muscle attachments.

Sexual selection A form of natural selection in which individuals compete with members of the same sex to mate with individuals of the opposite sex.

Single circulation The circulatory system of fish, where the heart pumps blood to the gills and onto the capillaries of the body tissues, before returning to the heart.

Somatic cells All of the cells of the body that are not involved in the formation of gametes.

Stabilizing selection Natural selection in which the population mean stabilizes on a particular non-extreme trait value, often due to conflicting selection pressures. The most common phenotype in the population is selected for and continues to dominate in future generations.

Supraorbital torus A bony ridge or thickening of the skull above the orbits.

Symbolic links Agreed-upon abstract, symbolic, verbal relationships that form the basis of human language.

Symbolic systems Behaviours which act as signs that represent meanings. Culture, writing, and talking are examples of symbolic systems.

Sympatric speciation when a new species evolves from the ancestral species while living in the same geographic location.

Systemic circulation The part of the cardiovascular system which transports oxygenated blood away from the left ventricle of the heart to the body tissues, and returns deoxygenated blood back to the right atrium of the heart.

Taphonomy Taphos, for 'death', and onomy for 'study of'. A science that looks at a fossil in terms of how the organism died and the processes acting on it until discovered.

Teleology A way of thinking that suggests evolution has intrinsic goals, and that characteristics develop in order to fulfil a purpose.

Temporal fossa A space formed between the temporal bone and the zygomatic arch. The temporalis passes through this space and the size of the fossa indicates the importance of the muscle in chewing.

Transcriptome The set of RNA molecules produced by active parts of the genome of cells or tissues.

Transmutation The conversion or transformation of one species into another.

Transposable elements A sequence of DNA that can change its position within a genome.

Transposon An alternative name for a transposable element.

Vertical gene transfer The transfer of genetic material from parents to offspring through sexual or asexual reproduction.

Wild type The phenotype of an organism that is typical (most common) in its natural environment. The term is often used in respect to laboratory or domestic organisms (eg applied to types of fruit fly, *Drosophila spp*).

INDEX